Behind the Scenes

Secrets of How Things Are Made

Authored by
Zahid Ameer

Published by

Goodword eBooks

Copyright © 2024 Zahid Ameer

All rights reserved.

ISBN: 9798343708165

DEDICATION

"I dedicate this book to my beloved parents, whose wisdom I hold in the highest regard. Their every word of guidance has been a beacon of light, illuminating the path of my life and shaping the essence of who I am."

Behind the Scenes

Contents:

Contents:

Introduction

> **The Power of Curiosity**
>
> **Bridging the Gap Between Complex Science and Everyday Life**
>
> **A World Built on Collaboration and Innovation**
>
> **Exploring a Range of Fields**
>
> **Why This Book Matters**
>
> **What You Will Gain**

How Is Space Travel Planned?

> **1. Defining Mission Objectives**
>
> **2. Developing a Mission Design**
>
> **3. Testing and Validation**
>
> **4. Coordination and Collaboration**
>
> **5. Launch Operations**
>
> **6. In-Flight Operations**

7. Return and Recovery

Conclusion

How is electricity generated?

1. Thermal Power Generation (Fossil Fuels)

2. Hydropower Generation

3. Nuclear Power Generation

4. Wind Power Generation

5. Solar Power Generation

6. Geothermal Power Generation

7. Biomass Power Generation

8. Tidal Power Generation

Conclusion

How is chocolate made?

1. Harvesting the Cocoa Pods

2. Fermenting the Beans

3. Drying the Beans

4. Roasting the Beans

Behind the Scenes

5. Cracking and Winnowing

6. Grinding and Refining

7. Mixing Ingredients

8. Conching

9. Tempering

10. Molding and Cooling

11. Packaging

Conclusion

How are skyscrapers built?

1. Design and Planning

2. Site Preparation and Excavation

3. Foundation Construction

4. Building the Structural Frame

5. Floor-by-Floor Assembly

6. Vertical Systems: Elevators, Plumbing, and Electrical Work

7. Safety Systems

Behind the Scenes

8. Cranes and Construction Equipment

9. Sustainability and Green Building Technologies

10. Final Stages: Interiors and Finishing Touches

11. Testing and Final Inspections

12. Opening and Occupation

Conclusion

How are vaccines developed?

1. Exploratory Stage

2. Preclinical Stage

3. Clinical Development

4. Regulatory Review and Approval

5. Manufacturing and Distribution

6. Post-Approval Monitoring (Phase 4 Trials)

Types of Vaccines Developed

Challenges in Vaccine Development

Conclusion

How is space exploration conducted?

Behind the Scenes

1. Planning and Mission Objectives

2. Design and Engineering of Spacecraft

3. Launch Systems

4. Traveling Through Space

5. Data Collection and Scientific Research

6. Human Space Exploration

7. Space Robotics

8. Challenges of Space Exploration

9. The Future of Space Exploration

Conclusion

How is gold mined and refined?

1. Gold Mining: Extracting the Ore

    a. Placer Mining:

    b. Hard Rock (Lode) Mining:

2. Extracting Gold from Ore:

    a. Cyanide Leaching:

    b. Mercury Amalgamation:

    c. Gravity Concentration:

3. Refining Gold:

    a. Miller Process:

    b. Wohlwill Process:

    c. Cupellation:

4. Smelting and Final Refining:

5. Post-Refinement:

Environmental and Ethical Considerations:

Conclusion:

How are diamonds formed?

    1. Carbon Source

    2. Depth and Pressure

    3. Crystallization Process

    4. Transport to the Surface

    5. Diamond Stability at the Surface

    6. Time Factor

    7. Lab-Grown Diamonds (Alternative Method)

Conclusion

How is artificial intelligence trained?

    1. Defining the Task

    2. Choosing the Right Algorithm

    3. Data Collection and Preprocessing

    4. Training the Model

    5. Validation and Tuning

    6. Testing the Model

    7. Deployment and Monitoring

    8. Transfer Learning

    9. Challenges in Training AI

Conclusion

How Are Weather Forecasts Predicted?

    1. Collecting Weather Data

    2. Inputting Data into Weather Models

    3. Running Simulations and Creating Forecasts

    4. Human Interpretation and Expertise

Behind the Scenes

5. Generating and Communicating the Forecast

6. Continuous Monitoring and Updates

The Role of Advanced Technologies

Why Weather Forecasting Matters

How is 3D Printing Done?

1. Creating a Digital Design

2. Slicing the Design

3. Setting Up the 3D Printer

4. The Printing Process

5. Post-Processing

6. Applications of 3D Printing

Conclusion

How are solar panels made?

1. Raw Materials: Silicon Extraction

2. Ingot Formation: Creating Silicon Blocks

3. Wafer Production: Slicing the Silicon Ingots

4. Doping: Creating a Semiconductor Layer

Behind the Scenes

5. Anti-Reflective Coating: Enhancing Efficiency

6. Metal Contacts: Collecting the Electric Current

7. Assembling the Solar Cells: Creating Solar Panels

8. Framing: Adding Durability

9. Testing and Quality Control

10. Installation and Operation

The Final Product: Solar Energy in Action

How is Cheese Produced?

Step 1: Milk Collection and Preparation

Step 2: Coagulation

Step 3: Cutting the Curds

Step 4: Cooking and Stirring

Step 5: Draining the Whey

Step 6: Salting

Step 7: Pressing (Optional)

Step 8: Aging (Affinage)

Step 9: Packaging and Distribution

Conclusion

How Are Movies Made from Script to Screen?

1. The Idea and Scriptwriting

2. Development and Pre-production

3. Production

4. Post-production

5. Distribution and Marketing

Conclusion

How Are Musical Instruments Made?

String Instruments (Violins, Guitars, Cellos)

Wind Instruments (Flutes, Clarinets, Trumpets)

Percussion Instruments (Drums, Marimbas)

Electronic Instruments (Synthesizers, Electric Guitars)

The Craftsmanship Behind the Music

How Is Space Debris Cleaned Up?

1. Laser Ablation

2. Harpoons and Nets

3. Robotic Arms and Tugging Satellites

4. Electrodynamic Tethers

5. Drag Sails

6. Space Sweepers

7. International Cooperation and Prevention

The Future of Space Cleanup

How is DNA Testing Done?

The Basics of DNA

Step 1: Collecting a DNA Sample

Step 2: Extracting the DNA

Step 3: Amplifying the DNA Using PCR

Step 4: Analyzing the DNA

Step 5: Interpreting the Results

Step 6: Reporting the Results

DNA Testing in Forensic Science

The Future of DNA Testing

Conclusion

How Are Drones Designed and Operated?

Designing a Drone: The Basics

Operating a Drone

Applications of Drones

The Future of Drones

How Is Glass Manufactured?

The Raw Materials

Melting: Turning Sand into Molten Glass

Shaping the Glass

Annealing: Cooling the Glass

Cutting and Finishing

Specialty Glass and Innovations

Recycling Glass

Conclusion

How Are Bridges Constructed?

Behind the Scenes

1. Planning and Design

2. Site Preparation

3. Foundation Construction

4. Superstructure Construction

5. Deck Installation

6. Finishing Touches and Safety Features

7. Opening and Maintenance

Conclusion

How is Data Encryption Done?

Understanding Data Encryption

The Importance of Encryption

How Encryption Works

1. Symmetric Encryption

How It Works:

2. Asymmetric Encryption

How It Works:

Additional Techniques in Data Encryption

Behind the Scenes

1. Hashing

2. Salting

3. Key Management

4. End-to-End Encryption (E2EE)

The Future of Data Encryption

Conclusion

How is Virtual Reality Created?

1. Understanding the Components of Virtual Reality

a. Hardware

b. Software

2. Designing the Virtual Environment

a. 3D Modeling and Animation:

b. Sound Design:

c. User Interface (UI) and User Experience (UX):

3. Developing the VR Experience

a. Prototyping:

    b. Programming:

    c. Testing:

4. Optimization and Finalization

    a. Performance Optimization:

    b. Final Touches:

5. Deployment and Distribution

    a. Publishing:

    b. User Support:

Conclusion

How is Weather Modified?

  Understanding Weather Modification

    What Is Weather Modification?

    The Science Behind Weather Modification

  Common Methods of Weather Modification

    1. Cloud Seeding

    2. Hail Suppression

    3. Fog Dispersal

4. Hurricane Modification

Applications of Weather Modification

Agricultural Benefits

Water Resource Management

Disaster Mitigation

Urban and Environmental Planning

Ethical Considerations and Challenges

1. Environmental Impact

2. Equity and Access

3. International Regulations

4. Public Perception

Conclusion

How Are Autonomous Vehicles Programmed?

1. Core Components of Autonomous Vehicles

2. Data Collection and Preparation

3. Machine Learning and AI Algorithms

4. Path Planning and Decision-Making

5. Control Systems Integration

6. Simulation and Testing

7. Safety and Regulatory Compliance

8. The Future of Autonomous Vehicle Programming

Conclusion

How is Coffee Processed from Bean to Cup?

1. Cultivation

2. Harvesting

3. Processing

4. Milling

5. Roasting

6. Grinding

7. Brewing

8. Enjoying the Final Product

Conclusion

How Are Volcanoes Studied?

Behind the Scenes

1. Field Studies and Observation

2. Remote Sensing Technologies

3. Geochemistry and Gas Emissions

4. Seismology and Ground Deformation

5. Volcanic Modeling and Simulation

6. Collaboration and Public Outreach

Conclusion

How is Currency Printed and Secured?

1. The Design Process

2. Material Selection

3. Printing Techniques

4. Adding Security Features

5. Quality Control

6. Distribution and Circulation

7. Counterfeit Prevention and Education

8. Continuous Improvement and Innovation

Conclusion

Behind the Scenes

How Are Smartphones Manufactured?

1. Design and Development

2. Sourcing Materials

3. Component Manufacturing

4. Assembly

5. Software Installation

6. Packaging and Shipping

7. After-Sales Support

Conclusion

How is Food Preserved?

The Importance of Food Preservation

Common Methods of Food Preservation

Conclusion

How is Oil Extracted and Refined?

The Extraction Process

1. Exploration

2. Drilling

3. Production

The Refining Process

    1. Distillation

    2. Cracking

    3. Reforming

    4. Treatment and Blending

Environmental Considerations

Conclusion

How Are Aircraft Designed and Tested?

    1. Conceptual Design

    2. Preliminary Design

    3. Detailed Design

    4. Prototyping

    5. Flight Testing

    6. Certification

    7. Production and Delivery

Conclusion

Behind the Scenes

How is the Internet Connected Globally?

1. **The Physical Infrastructure**

    a. **Undersea Cables**

    b. **Data Centers**

    c. **Satellite Networks**

2. **Internet Protocols and Standards**

    a. **Transmission Control Protocol/Internet Protocol (TCP/IP)**

    b. **Domain Name System (DNS)**

3. **Internet Service Providers (ISPs)**

    a. **Types of ISPs**

    b. **Peering Agreements**

4. **The Role of Government and Regulation**

    a. **Policy Development**

5. **The Future of Global Connectivity**

    a. **5G and Beyond**

    b. **The Digital Divide**

  Conclusion

How is the Human Genome Mapped?

  The Human Genome Project

  Steps in Mapping the Human Genome

  Advances Beyond the HGP

  Implications of Mapping the Human Genome

  Conclusion

How Are Robots Designed and Built?

  1. Understanding the Purpose and Requirements

  2. Conceptual Design

  3. Detailed Design and Prototyping

  4. Building the Robot

  5. Testing and Iteration

  6. Final Deployment and Maintenance

  7. Future Developments

  Conclusion

How is Renewable Energy Harnessed?

Behind the Scenes

Understanding Renewable Energy

Types of Renewable Energy Sources

The Process of Harnessing Renewable Energy

The Benefits of Renewable Energy

Challenges Ahead

Conclusion

How Are Satellites Launched into Orbit?

Understanding Satellites and Their Orbits

Types of Orbits

The Launch Process

1. Design and Development

2. Manufacturing

3. Launch Vehicle Selection

4. Pre-Launch Preparations

5. The Launch

6. Reaching Orbit

7. Post-Launch Operations

Conclusion

How Is Paper Recycled?

    1. Collection

    2. Sorting

    3. Shredding

    4. Pulping

    5. Cleaning

    6. Bleaching (Optional)

    7. Papermaking

    8. Finishing and Cutting

    9. Reuse

    Benefits of Paper Recycling

    Conclusion

How Are Fireworks Made?

    The Components of Fireworks

    The Manufacturing Process

        1. Ingredient Preparation

2. Mixing the Chemical Compositions

3. Creating the Stars

4. Assembling the Shells

5. Adding the Fuses

6. Final Assembly and Packaging

Safety Precautions

The Art of Pyrotechnics

Conclusion

How Are Pandemics Controlled?

1. Early Detection and Surveillance

2. Quarantine and Isolation Measures

3. Contact Tracing

4. Vaccination Campaigns

5. Public Health Communication

6. Pharmaceutical Interventions

7. Global Collaboration and Governance

8. Non-Pharmaceutical Interventions (NPIs)

9. Economic Support and Recovery Plans

Conclusion

How is a Supercomputer Built?

1. The Purpose and Design

2. The Building Blocks: Processors and Nodes

3. Memory and Storage

4. Cooling Systems

5. Power Supply and Energy Efficiency

6. Software and Programming

7. Testing and Optimization

8. Applications and Real-World Impact

Conclusion

How Are Video Games Developed?

1. Concept and Idea Development

2. Pre-Production

3. Production

4. Testing

5. Post-Production and Launch

6. Post-Launch Support

Conclusion

How is Water Purified for Drinking?

The Sources of Water

The Step-by-Step Process of Water Purification

Advanced Water Purification Techniques

Why Water Purification Is Vital

Conclusion: A Global Responsibility

How is Cryptocurrency Mined?

The Basics of Cryptocurrency Mining

How Mining Works: The Proof of Work Concept

The Role of Mining Hardware

Mining Farms and Energy Consumption

Difficulty Adjustments and Halving

Mining Pools: Strength in Numbers

Cloud Mining

Behind the Scenes

Challenges and Risks of Cryptocurrency Mining

Conclusion

How Are Ancient Artifacts Preserved?

1. Excavation and Initial Handling

2. Preventing Deterioration

3. Chemical Conservation Techniques

4. Repair and Reconstruction

5. Long-Term Storage and Monitoring

6. Documentation and Digital Preservation

7. Ethical Considerations

Conclusion

How is Waste Managed in Space?

The Problem of Space Waste

Waste Disposal Systems in Space

Space Debris: The Bigger Issue

Future Innovations in Space Waste Management

Conclusion

How Is Artificial Meat Grown in Labs?

The Science of Cultured Meat

Benefits of Lab-Grown Meat

Challenges and Future of Lab-Grown Meat

Conclusion

How Are Films Animated?

1. Concept and Storyboarding

2. Character Design and Model Sheets

3. Animatic Creation

4. Animation Techniques

2D Animation (Traditional Animation)

Stop-Motion Animation

3D CGI Animation

5. Rigging and Motion

6. Animating the Scenes

7. Lighting, Texturing, and Rendering

8. Sound Design and Voice Acting

9. Post-Production and Final Touches

Conclusion

How is Nuclear Energy Produced?

The Process of Nuclear Fission

The Role of the Reactor Core

Types of Nuclear Reactors

Safety Mechanisms and Concerns

Nuclear Energy: Benefits and Challenges

The Future of Nuclear Energy

How Is Virtual Money Regulated?

Understanding Virtual Money

The Need for Regulation

Current Regulatory Approaches

Challenges of Regulation

The Future of Virtual Money Regulation

Conclusion

Conclusion

- **The Hidden Complexity Behind Everyday Life**
- **The Power of Innovation and Human Ingenuity**
- **The Interconnected Nature of Our World**
- **Looking to the Future: Sustainability and the Role of Innovation**
- **The Journey Continues**
- **Final Thoughts**

Bibliography

Acknowledgments

Disclaimer

About me

Introduction

Welcome to "Behind the Scenes: Secrets of How Things Are Made", a journey into the hidden processes and intricate details of how the world around us operates. In an era where technology evolves at a dizzying pace and human innovation knows no bounds, we rarely stop to think about how the things we encounter every day are made, built, or brought into existence. From the food we eat, to the gadgets we use, to the monumental feats of engineering that shape our cities, there is a fascinating story behind each creation—a story of ingenuity, collaboration, and science.

We live in a world surrounded by astonishing creations, yet we often take for granted the processes behind them. Have you ever wondered how electricity flows from a power plant into your home, how vaccines are researched and developed to save lives, or how massive skyscrapers stand tall in the face of natural disasters? These questions—and countless others—are the ones this book seeks to explore.

This book opens the doors to the workshops, laboratories, factories, and creative minds where innovation happens. It pulls back the curtain on the industries, methods, and technologies that have revolutionized the world as we know it. In **"Behind the Scenes: Secrets of How Things**

Are Made," you'll find 100 captivating stories that reveal the mechanics, craftsmanship, and sheer brilliance required to turn an idea into reality.

The Power of Curiosity

Curiosity is a fundamental part of being human. It fuels our desire to understand the world and has led to some of the greatest discoveries in history. This book taps into that innate curiosity by presenting not just answers, but explanations that ignite a deeper understanding of the world around you. With every chapter, you'll gain insights into how seemingly simple objects—like your morning cup of coffee—are the result of complex, sometimes surprising processes. You'll learn about the incredible engineering that allows us to build underwater tunnels or send spacecraft to Mars. You'll also discover how ancient techniques have evolved into modern marvels, connecting the past with the present in ways that continue to shape our future.

Bridging the Gap Between Complex Science and Everyday Life

One of the key goals of this book is to demystify the technical aspects of how things are made and bring them to the everyday reader in a simple, engaging way. While some topics may seem daunting at first—like the inner workings of nuclear power plants or the intricate coding behind artificial intelligence—each subject is broken down into manageable, easy-to-understand concepts. Whether you have a background in science and technology or are simply a curious mind, this book has something to offer.

In **"Behind the Scenes,"** you will explore a diverse array of fields, from the industrial giants that power the global economy, to the niche technologies shaping the future of medicine, to the artistic techniques that breathe life into films and music. Each chapter stands as a gateway to deeper knowledge, blending technical explanations with captivating stories of human achievement.

A World Built on Collaboration and Innovation

Another central theme of this book is the idea that nothing in our world is created in isolation. Every process, no matter how complex or simple, relies on collaboration and innovation. Whether it's the teamwork behind constructing an awe-inspiring bridge or the synergy between scientists and researchers developing life-saving vaccines, **human ingenuity and collaboration** are at the heart of every story.

We'll explore how teams of engineers, architects, scientists, artists, and visionaries come together to solve problems, overcome obstacles, and create the everyday products and monumental achievements we often take for granted. The modern world is interconnected, and this

Behind the Scenes

book will highlight just how much each discipline—be it science, technology, art, or business—relies on the others.

Exploring a Range of Fields

"Behind the Scenes" takes you on a tour through a wide range of industries and fields. Here's a glimpse of the exciting areas we will explore:

- **Engineering Marvels:** Ever wondered how skyscrapers defy gravity, how underwater tunnels are constructed, or how massive dams control the flow of water? These feats of engineering showcase the brilliant minds behind the structural wonders that shape our modern world.
- **Technological Innovations:** Technology plays a central role in our daily lives, but how much do we really understand about the devices we use? From the design of smartphones to the advancements in artificial intelligence, this section reveals the inner workings of today's cutting-edge technology.
- **Natural Sciences and Medicine:** Our understanding of the natural world has led to the development of powerful tools in medicine, agriculture, and environmental management. Learn how scientists develop vaccines, genetically modify crops for sustainability, and even clone animals.

- **Art and Entertainment:** Not all creations are physical structures or technological advancements. We'll dive into the artistry behind filmmaking, animation, musical instrument production, and even how fireworks are meticulously crafted to create awe-inspiring displays.
- **Sustainability and the Future:** As the world faces environmental challenges, we look at the ways industries are adapting. How are solar panels manufactured? What are the methods used to clean up space debris? These are just some of the questions we'll explore as we look toward the future of sustainable living.

Why This Book Matters

In today's fast-paced world, it's easy to get lost in the rapid advances in technology, manufacturing, and science without truly understanding the mechanisms behind them. **"Behind the Scenes"** serves as a reminder that everything around us—from the simplest household item to the grandest feats of human achievement—begins with a process, a plan, and often years of development.

This book seeks to foster a sense of appreciation for the ingenuity and hard work involved in creating the world we live in. By learning how things are made, you gain a deeper connection to the products, services, and infrastructure that form the fabric of our society.

What You Will Gain

By the end of this book, you'll have:

- A clearer understanding of the processes behind the everyday objects and large-scale structures that surround you.
- Insights into the innovative minds and cutting-edge technologies that push the boundaries of what's possible.
- A newfound appreciation for the unseen efforts that go into creating the tools, products, and advancements that make our lives better.

So, join me on this exploration into the fascinating world of how things are made. Whether you're a curious individual, a tech enthusiast, or someone who simply loves learning new things, **"Behind the Scenes: Secrets of How Things Are Made"** will provide you with a fresh perspective on the world around you.

Let's dive into the details, one chapter at a time, and uncover the secrets that power the modern world!

Behind the Scenes

How Is Space Travel Planned?

Space travel is one of humanity's most ambitious endeavors, requiring meticulous planning, advanced technology, and seamless collaboration across multiple disciplines. The process of planning a space mission involves several key phases, each critical to ensuring the safety of the crew, the success of the mission, and the advancement of scientific knowledge. This exploration of how space travel is planned reveals the intricacies of aerospace engineering, logistics, science, and international cooperation.

1. Defining Mission Objectives

The first step in planning any space mission is defining its objectives. Objectives can vary widely, from conducting scientific research and technology demonstrations to exploring celestial bodies or supporting human habitation in space. For example, NASA's Artemis program aims to return humans to the Moon and establish a sustainable presence there, while missions like the Mars rover expeditions focus on scientific exploration and the search for extraterrestrial life.

Key considerations include:

- **Scientific Goals:** What research questions need to be answered? Are there specific experiments to conduct?
- **Technological Development:** What technologies need to be tested or developed? This could include life support systems, propulsion methods, or habitats.
- **Timeline and Funding:** How long will the mission take, and what budget is available? These elements will influence all subsequent planning.

2. Developing a Mission Design

Once objectives are established, the next step is to develop a comprehensive mission design. This phase involves creating a detailed blueprint for how the mission will be executed, including launch, trajectory, operations in space, and return.

Components of mission design include:

- **Launch Vehicle Selection:** Choosing the right rocket is crucial. Different rockets have varying payload capacities, thrust capabilities, and costs. For example, NASA's Space Launch System (SLS) is designed for deep space missions.
- **Trajectory Planning:** Calculating the spacecraft's trajectory involves complex mathematics and physics, ensuring the vehicle reaches its destination

efficiently. This includes calculating gravitational assists and optimizing fuel usage.
- **Spacecraft Design:** Engineers work on designing the spacecraft, considering factors like size, materials, and systems for navigation, communication, power, and life support.

3. Testing and Validation

Before launching a mission, extensive testing and validation are necessary to ensure the spacecraft and its systems will function correctly in the harsh environment of space.

Testing procedures include:

- **Simulation and Modeling:** Engineers use simulations to predict how the spacecraft will behave during launch, orbit, and re-entry. This includes modeling the spacecraft's systems and potential failure scenarios.
- **Prototype Testing:** Creating and testing prototypes allows engineers to validate designs. This could involve thermal vacuum tests, vibration tests, and more to mimic the conditions of space.
- **Crew Training:** If the mission involves human astronauts, rigorous training is essential. Astronauts practice in simulators, undergo physical training,

and learn emergency procedures to prepare for various scenarios.

4. Coordination and Collaboration

Space travel is rarely a solo endeavor; it often requires collaboration between various organizations, countries, and stakeholders. Coordination is key to a successful mission.

Elements of collaboration include:

- **International Partnerships:** Many space missions involve partnerships between countries, such as the International Space Station (ISS), which is a collaboration between NASA, Roscosmos, ESA, JAXA, and CSA.
- **Industry Collaboration:** Space agencies often work with private companies for the development of technology and launch services. Companies like SpaceX and Blue Origin play vital roles in modern space exploration.
- **Public Engagement:** Engaging the public and stakeholders helps build support for missions. Outreach programs can involve educational initiatives, live broadcasts of launches, and interactive platforms for public participation.

5. Launch Operations

With planning complete and systems validated, the mission transitions into the launch phase. This is a critical period where all the previous efforts come together.

Key aspects of launch operations include:

- **Launch Window:** Determining the optimal launch window is crucial, particularly for missions to other planets. Launch windows are influenced by the relative positions of celestial bodies.
- **Countdown Procedures:** A series of countdown procedures and checks are executed leading up to launch, ensuring all systems are functional and that the vehicle is ready for launch.
- **Launch Execution:** On launch day, teams work together to execute the launch, monitoring systems and managing any last-minute issues that may arise.

6. In-Flight Operations

Once the spacecraft is in orbit or on its way to its destination, mission control takes over, overseeing the flight and managing operations.

In-flight operations involve:

- **Monitoring Systems:** Continuous monitoring of spacecraft systems is vital to ensure everything is

functioning correctly. Any anomalies must be addressed promptly.
- **Data Collection:** During the mission, data is collected from instruments and experiments. This information is transmitted back to Earth for analysis.
- **Adaptability:** Mission control must be prepared to adapt to unexpected challenges, such as equipment malfunctions or changes in trajectory. This flexibility is crucial for mission success.

7. Return and Recovery

The final phase of a space mission is returning to Earth. This involves planning for re-entry, landing, and recovery of the spacecraft and crew.

Key considerations include:

- **Re-Entry Planning:** Re-entry involves navigating through the atmosphere at high speeds, requiring precise calculations to ensure a safe descent.
- **Landing Procedures:** Depending on the mission, landing could occur in the ocean (as with some NASA missions) or on land (like the SpaceX Crew Dragon). Recovery teams are prepared to assist once the spacecraft lands.
- **Post-Mission Analysis:** After recovery, a thorough analysis of the mission's data, systems, and overall

success takes place. This analysis informs future missions and advancements in technology.

Conclusion

Planning a space mission is an intricate, multi-step process that combines science, engineering, and international collaboration. From defining objectives to executing a successful launch and return, each phase requires careful consideration and coordination. As humanity continues to push the boundaries of space exploration, the lessons learned from these missions not only enhance our understanding of the universe but also drive innovation and inspire future generations to reach for the stars. Through **"How Is Space Travel Planned?"**, we gain insight into the meticulous efforts and brilliant minds that make the exploration of space possible.

Behind the Scenes

How is electricity generated?

Electricity generation is a fascinating and essential process that powers almost every aspect of modern life. Whether it's the lights in your home, the appliances you use, or the massive industrial systems that fuel economies, electricity is the invisible force behind it all. But how exactly is electricity generated? Let's explore the key methods and technologies used to produce electricity, from traditional sources like fossil fuels to renewable energy solutions.

1. Thermal Power Generation (Fossil Fuels)

One of the most common methods of generating electricity is through thermal power plants, which burn fossil fuels such as coal, natural gas, or oil. Here's how it works:

- **Fuel Combustion:** Fossil fuels are burned in a boiler to produce heat. The heat is used to convert water into high-pressure steam.
- **Turbine Rotation:** The steam flows through turbines, which are large, fan-like blades connected to a rotor. The force of the steam causes the turbines to spin at high speeds.
- **Generator:** The spinning turbines are connected to a generator, which consists of a rotating magnet inside a coil of copper wire. As the turbine spins the

magnet, it induces an electric current in the wire, generating electricity.
- **Condensation and Reuse:** After the steam passes through the turbines, it is cooled and condensed back into water to be reused in the cycle.

While thermal power plants have historically been the backbone of electricity generation, they rely on finite resources (fossil fuels) and emit greenhouse gases, contributing to climate change.

2. Hydropower Generation

Hydropower harnesses the energy of flowing water to generate electricity, often from rivers, dams, or water released from reservoirs. The process is simple yet effective:

- **Water Flow:** Water flows from a higher elevation (usually stored behind a dam) and is directed through turbines.
- **Turbine Spin:** The force of the moving water causes the turbines to spin, much like the steam does in a thermal power plant.
- **Electricity Generation:** The spinning turbines are connected to generators, which convert the mechanical energy into electricity.

Hydropower is a renewable source of energy and one of the oldest methods of electricity generation, used in many parts of the world due to its reliability and efficiency.

3. Nuclear Power Generation

Nuclear power plants generate electricity through a process called nuclear fission, where atoms of uranium or plutonium are split to release a massive amount of energy. Here's how nuclear power works:

- **Nuclear Reaction:** Inside the nuclear reactor, uranium atoms are bombarded with neutrons, causing them to split. This splitting releases a tremendous amount of heat energy.
- **Steam Production:** The heat produced from the nuclear reaction is used to convert water into steam, just like in a thermal power plant.
- **Turbine Rotation and Electricity Generation:** The steam drives turbines connected to generators, producing electricity.
- **Cooling and Condensation:** After the steam passes through the turbines, it is cooled, condensed back into water, and reused in the system.

Nuclear power plants produce no greenhouse gases during operation, making them a cleaner alternative to fossil fuels, but they generate radioactive waste that must be carefully managed.

Behind the Scenes

4. Wind Power Generation

Wind power is a form of renewable energy that captures the kinetic energy of wind and converts it into electricity using wind turbines. Here's how it works:

- **Wind Turbines:** Tall structures called wind turbines are placed in areas with strong, consistent winds (onshore or offshore). The wind turns the large blades of the turbine.
- **Turbine Rotation:** The rotation of the blades spins a rotor connected to a generator.
- **Electricity Generation:** As the rotor spins, it generates electricity by converting the mechanical energy of the moving air into electrical energy.

Wind power is one of the fastest-growing sources of renewable energy, producing no emissions or waste. However, it relies on weather conditions, so it cannot produce electricity consistently in all locations.

5. Solar Power Generation

Solar energy harnesses the power of the sun to generate electricity. There are two primary methods of solar power generation: photovoltaic (PV) cells and concentrated solar power (CSP).

- **Photovoltaic (PV) Cells:** Solar panels made up of many PV cells convert sunlight directly into electricity. When sunlight hits these cells, it excites electrons in the material, creating an electric current.
- **Concentrated Solar Power (CSP):** In CSP systems, mirrors are used to focus sunlight onto a small area, heating a fluid (often molten salt) to generate steam. The steam drives turbines connected to generators, producing electricity.

Solar power is a renewable and clean source of energy, but like wind power, it depends on environmental conditions and requires storage solutions for use during cloudy periods or nighttime.

6. Geothermal Power Generation

Geothermal power plants generate electricity by tapping into the Earth's natural heat. This heat, stored in rocks and fluids beneath the Earth's surface, can be used to produce electricity in areas with geothermal activity, such as near tectonic plate boundaries.

- **Hot Water Extraction:** Wells are drilled into the Earth's surface to access hot water or steam trapped underground.
- **Turbine Operation:** The steam is used to drive turbines, which are connected to generators that produce electricity.

- **Recycling Water:** After the steam cools and condenses back into water, it is often pumped back into the ground to be reheated and reused.

Geothermal energy is a reliable and continuous source of clean energy, but its availability is limited to regions with significant geothermal activity.

7. Biomass Power Generation

Biomass power involves burning organic materials such as wood, agricultural waste, and even landfill gas to generate electricity. Here's how biomass power generation works:

- **Combustion of Biomass:** Organic materials are burned in a boiler, generating heat.
- **Steam Production:** The heat from the combustion is used to produce steam, which drives turbines connected to a generator.
- **Electricity Generation:** As the turbines spin, the generator produces electricity.

Biomass is considered a renewable energy source as long as the organic material is sustainably managed. However, it can produce emissions, so the environmental impact depends on the methods used.

8. Tidal Power Generation

Behind the Scenes

Tidal power is generated by harnessing the energy from the rise and fall of tides. Tidal energy systems use the movement of seawater to spin turbines, which generate electricity.

- **Tidal Barrages or Tidal Streams:** Tidal barrages are dams built across tidal estuaries, and they capture the energy from the water flowing in and out with the tides. Tidal stream generators, on the other hand, capture the kinetic energy of moving water in tidal currents.
- **Turbine Movement:** As the water flows through the turbines, it causes them to spin.
- **Electricity Generation:** The turbines are connected to generators, which convert the mechanical energy into electricity.

Tidal power is renewable and predictable, as tidal cycles are consistent, but the technology is still in its early stages and requires suitable coastal locations.

Conclusion

Electricity generation is a diverse and complex process involving a wide range of methods and technologies. From burning fossil fuels to harnessing the power of the sun, wind, and water, each method plays a critical role in meeting the world's energy needs. While traditional methods like coal and natural gas have been the dominant

Behind the Scenes

sources of electricity, the future is leaning more toward renewable energy sources like wind, solar, and hydropower as we seek to reduce carbon emissions and combat climate change.

As technology continues to advance, we can expect even more innovative ways to generate electricity, making it cleaner, more efficient, and more sustainable for generations to come.

How is chocolate made?

The process of making chocolate is a fascinating journey that transforms raw cocoa beans into the delicious, smooth, and flavorful treat loved by millions. Here's a step-by-step look at how chocolate is made, from bean to bar:

1. Harvesting the Cocoa Pods

Chocolate begins its life as a seed inside a fruit, the cocoa pod, which grows on the cacao tree (Theobroma cacao). These trees are primarily grown in tropical regions near the equator, with countries like Ivory Coast, Ghana, and Indonesia leading global production. The cocoa pods are harvested when they ripen, changing color from green to yellow or orange. Farmers use machetes to cut the pods from the trees carefully.

2. Fermenting the Beans

After harvesting, the cocoa pods are opened, and the beans, covered in a sweet, sticky pulp, are removed. The beans are then placed in shallow containers, where they undergo fermentation for about 5 to 7 days. Fermentation is crucial as it develops the beans' flavor and removes much of the bitterness. The beans are turned regularly to ensure even fermentation, and during this process, their purple color gradually turns brown.

3. Drying the Beans

Once fermentation is complete, the cocoa beans are spread out under the sun to dry for several days. Drying reduces the moisture content in the beans, making them easier to transport and preventing mold growth. Farmers rake the beans frequently to ensure uniform drying. At this stage, the beans develop their signature chocolatey aroma. Once dry, the beans are packed and shipped to chocolate manufacturers around the world.

4. Roasting the Beans

At the chocolate factory, the dried cocoa beans are roasted to enhance their flavor further. The roasting process involves heating the beans to a temperature between 250°F to 350°F (120°C to 180°C) for 30 minutes to 2 hours, depending on the desired flavor profile. The roasting process brings out the rich, chocolatey notes while removing any remaining bitterness. It also loosens the outer shells of the beans, making them easier to crack.

5. Cracking and Winnowing

After roasting, the cocoa beans are cracked open to separate the outer shell from the inner nib. The nibs are the edible part of the cocoa bean, rich in cocoa butter and flavor. This process, known as winnowing, removes the shell and leaves behind the nibs, which are then ground into a liquid form.

6. Grinding and Refining

The roasted nibs are then ground into a thick paste known as **cocoa mass** or **cocoa liquor** (despite the name, it contains no alcohol). During grinding, the heat generated causes the cocoa butter inside the nibs to melt, creating a liquid. This cocoa liquor forms the base for all chocolate products, whether it's dark, milk, or white chocolate. At this stage, the cocoa mass can be processed further into cocoa powder and cocoa butter, two essential components of chocolate-making.

7. Mixing Ingredients

To make chocolate, the cocoa mass is mixed with other ingredients, depending on the type of chocolate being produced:

- **Dark Chocolate:** Cocoa mass, sugar, and sometimes additional cocoa butter.
- **Milk Chocolate:** Cocoa mass, sugar, cocoa butter, and milk powder.
- **White Chocolate:** Cocoa butter, sugar, and milk powder (no cocoa solids).

This mixture is blended and refined until smooth, with any large cocoa particles broken down to create a uniform, creamy texture.

8. Conching

The chocolate mixture then goes through a process called **conching**, where it is continuously mixed and aerated at controlled temperatures for several hours or even days. This step smooths out the chocolate and allows volatile acids and undesirable flavors to evaporate, enhancing the final taste. The longer the conching, the smoother the chocolate becomes, and the more refined its flavor.

9. Tempering

Tempering is the process of carefully cooling and reheating the chocolate to stabilize the cocoa butter crystals, giving the finished product a glossy shine and smooth texture. Properly tempered chocolate also has a satisfying "snap" when broken. This step is critical for ensuring the chocolate has the right consistency and melts evenly when eaten.

10. Molding and Cooling

Once tempered, the liquid chocolate is poured into molds to create chocolate bars, truffles, or other shapes. The molds are then placed in a cooling tunnel or refrigerator to harden the chocolate quickly. Once fully cooled and solidified, the chocolate can be removed from the molds and packaged.

11. Packaging

The final step in the chocolate-making process is packaging. Whether wrapped in foil, paper, or plastic, the chocolate is sealed to maintain its freshness and protect it from moisture, air, and light, which can affect its flavor and texture. The chocolate is now ready for distribution to stores and into the hands of eager consumers!

Conclusion

From the tropical cacao farms to the meticulous processes inside chocolate factories, the journey of chocolate is an intricate blend of nature, science, and artistry. Each step, from fermentation to tempering, plays a critical role in developing the rich flavors and smooth textures that make chocolate the universally beloved treat it is today. Whether dark, milk, or white, the magic behind chocolate lies in the delicate transformation of cocoa beans into something irresistible.

Behind the Scenes

How are skyscrapers built?

Building skyscrapers is a remarkable engineering achievement that requires precise planning, advanced technology, and a deep understanding of physics and materials. The process involves various phases, from design and site preparation to construction and installation of systems. Let's take a detailed look at how skyscrapers are built.

1. Design and Planning

Before any construction begins, architects and engineers work together to design the skyscraper. This process includes:

- **Architectural Design**: The building's overall look, layout, and functionality are planned by architects. They consider factors like space usage, aesthetics, and environmental sustainability.
- **Structural Engineering**: Engineers ensure the building will be strong enough to withstand forces like gravity, wind, and potential earthquakes. They choose materials like steel or reinforced concrete to support the structure.
- **Wind and Load Analysis**: One of the key challenges in skyscraper construction is the effect of wind. Engineers analyze how wind will affect the

building's stability and design it to sway slightly without compromising safety.
- **Building Codes and Regulations**: Skyscraper designs must comply with local regulations and building codes, covering fire safety, accessibility, and structural integrity.

2. Site Preparation and Excavation

Once the design is finalized and approved, the construction site is prepared. This phase includes:

- **Clearing the Site**: The construction site must be cleared of any existing structures, trees, or debris. In urban areas, this can be complex, especially if the site is surrounded by other buildings.
- **Excavation**: Skyscrapers require deep, solid foundations, so excavation work begins by digging down into the earth. The depth can vary depending on the height and size of the building. For taller skyscrapers, basements and underground parking may also be constructed.
- **Shoring**: During excavation, shoring systems are often used to prevent soil collapse and maintain the stability of surrounding structures.

3. Foundation Construction

The foundation is critical for a skyscraper's stability. There are two common types of foundations used:

- **Pile Foundations**: Long columns, or piles, are driven deep into the ground, often down to bedrock. These piles help support the immense weight of the building. The piles can be made of steel, concrete, or wood.
- **Mat Foundations**: A thick concrete slab, known as a mat, is poured over the entire footprint of the building. It helps evenly distribute the weight of the skyscraper over a large area of the ground.

4. Building the Structural Frame

Once the foundation is in place, construction of the skyscraper's frame begins. This is the skeleton that will support the building:

- **Steel Framework**: Modern skyscrapers typically use a steel frame. Steel beams and columns are bolted or welded together to form the building's skeleton. Steel is both strong and flexible, allowing it to bend slightly without breaking, which is vital for wind resistance and earthquake protection.
- **Concrete Construction**: Some skyscrapers use reinforced concrete instead of steel. Concrete is extremely strong in compression and is often used

for the core of the building, which houses elevators and stairwells.
- **Core Construction**: The core, often built first, provides lateral stability and houses essential systems like elevators, utilities, and emergency stairways. It's usually made from reinforced concrete.

5. Floor-by-Floor Assembly

With the structural frame rising, floors are constructed level by level:

- **Floor Installation**: Prefabricated floor panels, often made of concrete or steel, are hoisted by cranes and installed on each level. These panels are bolted or welded to the steel framework.
- **External Cladding**: As floors go up, the building's exterior, or cladding, is added. This can be glass, steel, or other materials that offer insulation and protection from the elements. Glass curtain walls are common in modern skyscrapers, giving them a sleek appearance.
- **Curtain Walls**: These walls are non-structural and hang from the building's frame. They provide protection from the weather and are often made of glass and metal panels.

Behind the Scenes

6. Vertical Systems: Elevators, Plumbing, and Electrical Work

Skyscrapers need advanced internal systems to function:

- **Elevators**: Modern skyscrapers use high-speed elevators to transport people efficiently between floors. Multiple elevator banks are installed, with some elevators designated for express service to higher floors. Smart elevators are now common, using algorithms to reduce wait times and energy consumption.
- **Plumbing and HVAC Systems**: Plumbing systems for water supply and sewage are installed alongside heating, ventilation, and air conditioning (HVAC) systems. HVAC systems are designed to maintain comfortable temperatures and air quality on all floors.
- **Electrical and Communication Systems**: Electrical wiring, internet cables, and other communication systems are threaded through the building to ensure a seamless supply of power and connectivity.

7. Safety Systems

Safety is a top priority in skyscraper construction, especially regarding fire and evacuation protocols:

- **Fireproofing**: Fire-resistant materials are used throughout the building. Fireproofing is applied to steel beams to prevent them from collapsing in the event of a fire.
- **Sprinkler Systems**: Automatic sprinkler systems are installed on each floor to contain and extinguish fires.
- **Stairwells and Exits**: Multiple stairwells are built to ensure that occupants can safely evacuate. In many skyscrapers, fire-resistant doors and pressurized stairwells prevent smoke from entering these escape routes.

8. Cranes and Construction Equipment

Cranes play a vital role in skyscraper construction:

- **Tower Cranes**: These massive cranes, often anchored to the building's core, lift heavy materials like steel beams and prefabricated sections to the upper floors. As the skyscraper grows, the cranes themselves are raised higher.
- **Construction Elevators**: Temporary construction elevators, called hoists, are installed to transport workers and materials to the higher floors as the skyscraper rises.

9. Sustainability and Green Building Technologies

Modern skyscrapers incorporate eco-friendly technologies to minimize environmental impact:

- **Energy-Efficient Systems**: Advanced HVAC systems, energy-efficient lighting, and smart building management systems are installed to reduce energy consumption.
- **Green Roofs and Renewable Energy**: Some skyscrapers have green roofs with plants to improve insulation and air quality. Solar panels and wind turbines may also be integrated to provide renewable energy.
- **LEED Certification**: Many skyscrapers are designed to meet LEED (Leadership in Energy and Environmental Design) certification standards, ensuring that the building is environmentally friendly in its construction and operation.

10. Final Stages: Interiors and Finishing Touches

As the skyscraper nears completion, the focus shifts to interior work and finishing:

- **Interior Finishes**: Walls, flooring, ceilings, and fixtures are installed inside the building. Interior designers often collaborate with architects to create aesthetically pleasing spaces.

- **Landscaping**: The area around the skyscraper is landscaped, with sidewalks, parks, and public spaces incorporated into the design.

11. Testing and Final Inspections

Before opening, the skyscraper must undergo rigorous testing:

- **Structural Tests**: Engineers conduct tests to ensure the building's structural integrity, including wind and earthquake simulations.
- **System Tests**: Electrical, plumbing, HVAC, and fire safety systems are thoroughly tested to ensure they meet building codes and safety standards.

12. Opening and Occupation

Once all inspections are completed, the skyscraper is ready to be occupied. Depending on its purpose, it may serve as office space, residential apartments, or mixed-use facilities. Tenants move in, and the building becomes part of the city's skyline, a testament to modern engineering and design.

Conclusion

Building a skyscraper is a complex process that requires cutting-edge technology, meticulous planning, and highly skilled workers. From laying the foundation to installing elevators, every step is a marvel of engineering. Skyscrapers not only symbolize human progress and architectural achievement, but they also demonstrate our ability to push the limits of what is possible, reaching ever closer to the sky.

Behind the Scenes

How are vaccines developed?

Vaccine development is a complex and multi-step process that involves years of research, testing, and evaluation to ensure both efficacy and safety. Vaccines are designed to protect individuals from infectious diseases by training the immune system to recognize and fight off harmful pathogens. Here's a detailed breakdown of how vaccines are developed:

1. Exploratory Stage

The vaccine development process begins in the **exploratory stage**, where scientists study the biology of the disease-causing virus or bacteria. This involves identifying which part of the pathogen could trigger an immune response. These parts could be proteins, sugars, or other molecular structures on the surface of the pathogen. During this stage, scientists also analyze how the immune system naturally fights the infection.

For example, if researchers are developing a vaccine for a virus, they might identify specific viral proteins that can be used as antigens (substances that provoke an immune response). This research is often conducted in laboratories using tools like genetic sequencing and bioinformatics.

2. Preclinical Stage

Behind the Scenes

Once researchers have identified potential antigens, the vaccine enters the **preclinical stage**. At this point, scientists create a vaccine candidate (a potential formulation of the vaccine) and test it in cell cultures and animals to assess its safety and ability to induce an immune response.

- **Animal Testing:** In this phase, the vaccine candidate is usually tested on small animals like mice or rabbits and then larger animals such as monkeys. Researchers observe whether the vaccine generates a sufficient immune response and whether there are any harmful side effects.

This stage provides valuable information about the immune response and helps researchers decide whether the vaccine candidate should move to human trials. However, not all vaccines move forward from this stage—many candidates fail due to safety concerns or insufficient immune response in animal models.

3. Clinical Development

If the vaccine candidate shows promise during the preclinical stage, it advances to **clinical trials** in humans, which are conducted in three phases.

- **Phase 1 Trials:** These trials involve a small group of healthy volunteers (usually 20 to 100 people).

The primary goal is to evaluate the vaccine's safety and determine the appropriate dosage. Researchers closely monitor participants for any adverse reactions and gather data on the immune response triggered by the vaccine.

- **Phase 2 Trials:** Once the vaccine is deemed safe in Phase 1, it is tested on a larger group of people (several hundred) to further assess its safety, immune response, and effectiveness. In Phase 2, participants may include individuals who are at higher risk of contracting the disease. Researchers also look at optimal dosing and delivery methods (injection, nasal spray, etc.).
- **Phase 3 Trials:** In this phase, the vaccine is tested on thousands or even tens of thousands of people to assess its efficacy and further evaluate its safety. These large-scale trials are usually conducted in different locations and among diverse populations. The goal is to confirm whether the vaccine effectively prevents infection in a real-world setting. Researchers also monitor for rare side effects that might not have appeared in earlier phases.

This stage can take several years to complete, as it requires the participation of many volunteers and careful monitoring of outcomes. In the case of emergency situations, such as during pandemics, regulatory agencies may allow for accelerated timelines.

4. Regulatory Review and Approval

After Phase 3 trials are successfully completed, the vaccine manufacturer submits an application to regulatory authorities, such as the U.S. **Food and Drug Administration (FDA)** or the **European Medicines Agency (EMA)**, for approval.

- **Review Process:** Regulatory agencies carefully review all the data from the clinical trials, including safety and efficacy data, the manufacturing process, and any potential side effects. Independent experts may also evaluate the findings. If the vaccine meets safety and efficacy standards, it is approved for public use.

In some cases, especially during global health emergencies (like the COVID-19 pandemic), regulatory bodies may issue an **Emergency Use Authorization (EUA)**. This allows the vaccine to be distributed and administered while still collecting additional data.

5. Manufacturing and Distribution

Once approved, vaccines enter the **manufacturing** phase, where they are produced on a large scale. Manufacturing vaccines is a highly specialized and rigorous process that follows strict standards to ensure consistency and quality.

- **Quality Control:** During production, vaccines must pass rigorous quality control tests to ensure that each batch meets the required standards for safety, potency, and purity.
- **Cold Chain Logistics:** Many vaccines need to be stored and transported at specific temperatures (known as the "cold chain"). For example, some vaccines require freezing or refrigeration to remain effective. Failure to maintain the cold chain can render vaccines ineffective.

6. Post-Approval Monitoring (Phase 4 Trials)

Even after a vaccine is approved and distributed, it continues to be monitored for **long-term safety and effectiveness** through **Phase 4 trials**. These trials involve ongoing surveillance of vaccinated individuals to detect any rare or long-term side effects that might not have appeared during earlier clinical trials.

- **Vaccine Adverse Event Reporting Systems (VAERS):** In countries like the U.S., healthcare providers, manufacturers, and the public can report any adverse events through systems like VAERS. This allows researchers to identify any previously unrecognized risks associated with the vaccine.

Types of Vaccines Developed

Behind the Scenes

Vaccines can be made using various approaches depending on the nature of the disease and the immune response required:

- **Inactivated Vaccines:** These vaccines use a killed version of the virus or bacteria to trigger an immune response (e.g., polio vaccine).
- **Live Attenuated Vaccines:** These vaccines use a weakened form of the pathogen, strong enough to provoke an immune response without causing the disease (e.g., measles, mumps, rubella [MMR] vaccine).
- **Subunit, Recombinant, or Conjugate Vaccines:** These vaccines use specific parts of the pathogen (like proteins or sugars) to stimulate the immune system (e.g., human papillomavirus [HPV] vaccine).
- **mRNA Vaccines:** These newer vaccines use messenger RNA to instruct cells to produce a protein from the pathogen, prompting an immune response (e.g., COVID-19 vaccines like Pfizer and Moderna).
- **Viral Vector Vaccines:** These vaccines use a modified virus to deliver genetic material that encodes antigens, inducing immunity (e.g., Johnson & Johnson's COVID-19 vaccine).

Challenges in Vaccine Development

The process of developing vaccines is filled with challenges, including:

- **Time and Cost:** Developing a vaccine can take anywhere from 10 to 15 years and cost hundreds of millions of dollars.
- **Mutating Pathogens:** Some pathogens, like influenza or HIV, mutate rapidly, making it difficult to develop a long-lasting vaccine.
- **Distribution:** Ensuring equitable global access to vaccines, especially in low-income countries, is another major hurdle.

Conclusion

Vaccine development is a remarkable scientific achievement that combines biology, technology, and medicine. The process from lab research to global distribution is long, complex, and requires the collaboration of scientists, regulators, and manufacturers. But once successful, vaccines become one of the most powerful tools in protecting human health and preventing disease outbreaks, safeguarding millions of lives around the world.

Behind the Scenes

How is space exploration conducted?

Space exploration is a complex and multi-disciplinary endeavor that involves a combination of advanced technology, rigorous scientific research, and international collaboration. Here's an overview of the key steps and processes involved in conducting space exploration:

1. Planning and Mission Objectives

Every space exploration mission begins with detailed planning and defining objectives. These objectives can vary from studying distant planets, searching for life on other celestial bodies, exploring asteroids, or understanding the effects of space on human physiology. The mission objectives guide the type of spacecraft needed, the scientific instruments to be used, and the overall mission timeline.

Agencies like NASA, ESA (European Space Agency), and private companies like SpaceX collaborate with scientists, engineers, and policymakers to establish clear goals. This can include studying Mars, observing galaxies, or launching telescopes like the Hubble Space Telescope.

2. Design and Engineering of Spacecraft

Behind the Scenes

Once objectives are set, engineers and scientists design a spacecraft or space system tailored to the mission. This involves:

- **Choosing the type of spacecraft:** It could be a satellite, rover, probe, or manned space vehicle.
- **Power systems:** Spacecraft are usually powered by solar panels, nuclear energy, or batteries.
- **Propulsion systems:** Rockets are used to launch spacecraft into space, while smaller thrusters control movement and orientation in space.
- **Scientific instruments:** Depending on the mission, instruments like cameras, spectrometers, radiation detectors, and seismometers are included for data collection.

The spacecraft must also be equipped with communication systems to send data back to Earth, as well as thermal protection systems to handle extreme temperatures in space.

3. Launch Systems

Launching a spacecraft into space requires powerful rockets that can overcome Earth's gravitational pull. The launch process includes:

- **Pre-launch testing:** Extensive tests are conducted to ensure that the spacecraft and all its systems are

Behind the Scenes

functioning properly. Simulations of space conditions, such as vacuum and temperature extremes, are also run.

- **Rocket selection:** Depending on the mission's size, weight, and distance (whether it's going to low-Earth orbit or deep space), different rockets are selected. For example, NASA's **Space Launch System (SLS)** is designed for deep space exploration, while SpaceX's **Falcon 9** is used for lower-Earth orbit missions.
- **Launch window:** A specific time frame, known as the launch window, is selected based on the alignment of the spacecraft's destination (e.g., Mars or the Moon) and the Earth's position. This ensures the most efficient and cost-effective trajectory.

4. Traveling Through Space

Once launched, spacecraft begin their journey toward their destination. Space travel involves:

- **Trajectory and navigation:** Spacecraft follow carefully calculated paths (trajectories) to reach their destination. They may use gravitational assists from planets to save fuel and gain speed, a process known as a slingshot maneuver.
- **Orbital insertion:** For missions to planets or moons, the spacecraft must perform precise

maneuvers to enter orbit around the target body. Failing to achieve this can result in missing the destination entirely.
- **Mid-course corrections:** Along the way, small adjustments are made to the spacecraft's trajectory using its thrusters to ensure it stays on course.

Space missions often take months or years, especially when exploring distant planets or outer reaches of the solar system, like with NASA's Voyager probes.

5. Data Collection and Scientific Research

Once the spacecraft reaches its destination, it begins conducting the scientific research for which it was designed. This involves:

- **Orbiters and probes:** Orbiters circle the planet or moon, collecting data on the atmosphere, surface, and magnetic field. Probes might land on the surface to take direct samples or transmit data.
- **Rovers:** In missions to Mars and the Moon, rovers like NASA's **Curiosity** or **Perseverance** are deployed to explore the terrain, analyze soil samples, and search for signs of water or life.
- **Telescopes:** Space-based telescopes like **Hubble** or **James Webb Space Telescope (JWST)** study distant galaxies, stars, and exoplanets, gathering data on their formation and evolution.

These instruments send data back to Earth via long-range communication systems, often using Deep Space Network (DSN) antennas to receive signals across vast distances.

6. Human Space Exploration

When humans are involved, space exploration becomes even more complex. Key aspects include:

- **Spacecraft for humans:** Manned missions require spacecraft like **NASA's Orion** or **SpaceX's Crew Dragon** that support life with air, water, food, and protection from radiation.
- **International Space Station (ISS):** The ISS serves as a hub for human space exploration, where astronauts live for extended periods to conduct experiments and prepare for future deep-space missions. The ISS orbits the Earth and is used for experiments in microgravity.
- **Space travel hazards:** Human missions face unique challenges such as radiation exposure, the effects of zero gravity on the body, and psychological stress. Long-term missions to Mars or beyond require solutions to these risks, like radiation shields and artificial gravity.
- **Moon and Mars missions:** As part of future space exploration, NASA's **Artemis program** aims to return humans to the Moon and later send astronauts

to Mars. This will involve advanced spacecraft and technologies that can support life on the Moon's surface or during interplanetary travel.

7. Space Robotics

Robots play an essential role in space exploration, performing tasks that would be dangerous or impossible for humans. Some notable uses of robots in space include:

- **Rovers and landers** for planetary exploration, such as Mars rovers.
- **Space drones** and robotic arms for maintenance and construction on the ISS.
- **Autonomous spacecraft** that can navigate and conduct missions without direct human control.

Robotics will be critical in future missions to build habitats on the Moon and Mars.

8. Challenges of Space Exploration

Space exploration presents a series of challenges that need to be overcome:

- **Extreme conditions:** Temperatures in space vary dramatically, from intense heat in the Sun's direct light to freezing cold in its absence. Spacecraft and instruments must be built to endure these extremes.

- **Radiation:** Spacecraft and astronauts are exposed to high levels of cosmic radiation, which can damage electronic components and human cells. Protective shielding and radiation-resistant materials are vital.
- **Distance and communication delays:** Space is vast, and as spacecraft venture farther from Earth, communication delays increase. For example, signals to and from Mars take about 20 minutes each way. Autonomous decision-making systems help spacecraft navigate when direct control is impossible.
- **Cost:** Space exploration is expensive. Developing spacecraft, conducting research, and ensuring safety involve enormous investments. However, collaborations between nations and private companies, such as SpaceX or Blue Origin, help distribute costs and accelerate technological advancements.

9. The Future of Space Exploration

The future of space exploration is brimming with potential and innovation. Some exciting developments include:

- **Mars colonization:** Missions like SpaceX's **Starship** aim to land humans on Mars and eventually build self-sustaining colonies.

- **Asteroid mining:** Future missions may target asteroids for mining valuable resources like water, metals, and minerals.
- **Interstellar exploration:** With missions like **NASA's Breakthrough Starshot**, the goal is to send tiny spacecraft to neighboring star systems to explore exoplanets and potentially find life.
- **Private space tourism:** Companies like **Blue Origin** and **Virgin Galactic** are pioneering space tourism, making space travel accessible to civilians.

Conclusion

Space exploration is an ongoing human endeavor that pushes the boundaries of what is possible. It merges advanced technology, scientific discovery, and human curiosity to explore the unknown. As we continue to venture deeper into space, from the Moon to Mars and beyond, the future promises even more groundbreaking discoveries that will redefine our understanding of the universe.

Behind the Scenes

How is gold mined and refined?

Gold mining and refining is a complex process that involves extracting gold from the earth and transforming it into its pure, final form. Let's break down the major steps involved:

1. Gold Mining: Extracting the Ore

Gold can be found in both placer deposits (where it has been eroded and washed down into rivers or streams) and lode deposits (where it is found in solid rock).

a. Placer Mining:

This method involves collecting gold from riverbeds, streambeds, and other deposits of sand and gravel where gold particles settle. The gold is extracted through techniques like panning, sluicing, or dredging:

- **Panning**: Involves using a pan to scoop up water and gravel. The gold, being denser, sinks to the bottom, allowing the miner to separate it from lighter materials.
- **Sluicing**: A sluice box is used to trap gold particles as water flows through. Gold gets caught in ridges

or mats in the sluice box while lighter materials are washed away.
- **Dredging**: A more industrial process where machines are used to vacuum gold-rich sediments from riverbeds.

b. Hard Rock (Lode) Mining:

In hard rock mining, gold is found in solid rock formations. The process includes:

- **Drilling and Blasting**: Miners drill into the rock and use explosives to break it apart.
- **Crushing**: The broken rocks are transported to a mill where they are crushed into smaller pieces to extract the gold ore.
- **Grinding**: The crushed rock is further ground to a fine powder in mills to separate gold from other minerals.

2. Extracting Gold from Ore:

Once the ore is mined, the gold must be extracted from other minerals and materials.

a. Cyanide Leaching:

A common method for low-grade ores is to use a chemical solution (typically sodium cyanide) to dissolve the gold:

- The ore is placed in large vats or heaps, and a cyanide solution is sprayed or poured over it.
- The cyanide bonds with the gold particles, dissolving them into the solution.
- The solution is then collected and processed to extract the gold.

b. Mercury Amalgamation:

Historically, mercury was used to extract gold from ore. Mercury forms an amalgam with gold, which can then be heated to evaporate the mercury, leaving behind pure gold. However, this method is now largely abandoned due to mercury's toxic environmental effects.

c. Gravity Concentration:

For ores with a high concentration of gold, gravity-based separation methods are used. Centrifugal force or shaking tables are employed to separate heavier gold particles from lighter material.

3. Refining Gold:

Once the raw gold has been extracted, it undergoes various refining processes to remove impurities and produce pure gold.

a. Miller Process:

In the Miller process, the gold is melted in a furnace, and chlorine gas is bubbled through it. The chlorine reacts with impurities, forming chlorides, which rise to the surface and can be skimmed off, leaving behind gold that is about 99.5% pure.

b. Wohlwill Process:

For even higher purity, the Wohlwill process is used. This method involves electrolysis, where impure gold is cast into an anode, and pure gold collects on a cathode when electricity is passed through a bath of gold chloride solution. This produces gold with a purity of up to 99.999%.

c. Cupellation:

An ancient refining method, cupellation involves heating the gold with lead in a furnace. The lead binds to impurities, leaving pure gold behind. This method is not commonly used in modern industrial refining but is historically significant.

4. Smelting and Final Refining:

The final step involves heating the refined gold to very high temperatures (around 1,064°C or 1,947°F), where it is melted and poured into molds to create bars, ingots, or other desired forms. Any remaining impurities are removed during the melting process.

5. Post-Refinement:

After the refining process, the gold can be further alloyed with other metals (such as copper or silver) to create gold products like jewelry, coins, or electronic components.

Environmental and Ethical Considerations:

Gold mining and refining have significant environmental and ethical impacts. Cyanide and mercury can cause pollution, and the mining process can disrupt ecosystems, leading to deforestation, water contamination, and loss of biodiversity. In response, many companies are adopting more sustainable practices, such as using eco-friendly chemicals or implementing better waste management systems. Additionally, concerns about human rights abuses and unethical labor practices in gold mining have led to efforts to promote "responsible" or "fair trade" gold.

Conclusion:

Gold mining and refining is a multi-stage process that begins with the extraction of ore from the earth and ends with the production of highly purified gold. Each step involves sophisticated techniques, chemistry, and machinery, and efforts are being made to minimize the environmental footprint of this valuable resource. By understanding the complexity of gold production, we can

Behind the Scenes

better appreciate its worth and the impact it has on both people and the planet.

Behind the Scenes

How are diamonds formed?

Diamonds, often seen as symbols of luxury and eternal love, are formed through a fascinating natural process that occurs deep beneath the Earth's surface. Their creation is a remarkable journey of immense pressure, extreme heat, and millions, if not billions, of years of geological activity. Here's a breakdown of how diamonds are formed:

1. Carbon Source

Diamonds are made of carbon, one of the most common elements on Earth. However, not just any carbon can form diamonds. The carbon atoms that become diamonds typically originate deep within the Earth's mantle, where unique conditions exist to transform ordinary carbon into this rare and valuable gemstone.

2. Depth and Pressure

The key to diamond formation is the immense **pressure and heat** found deep within the Earth, about 90 to 150 miles (140 to 240 kilometers) below the surface. At these depths, the Earth's temperature ranges from 1,650 to 2,370°F (900 to 1,300°C), and the pressure is about 725,000 pounds per square inch (5 gigapascals). These extreme conditions force the carbon atoms to bond in a highly ordered, crystalline structure, resulting in a diamond's characteristic hardness.

3. Crystallization Process

Under these intense conditions, carbon atoms bond together in a lattice structure, creating what is known as a **diamond cubic crystal system**. This crystal lattice is what makes diamonds the hardest naturally occurring substance on Earth. Unlike graphite, which is also made of carbon but has a softer structure, diamond's atoms are arranged in a tetrahedral structure, meaning each carbon atom is strongly bonded to four others in a three-dimensional framework.

4. Transport to the Surface

After diamonds form deep within the mantle, they are brought closer to the Earth's surface through powerful volcanic eruptions. These eruptions push magma through deep fractures in the Earth's crust, creating what are known as **kimberlite pipes**—vertical structures that act as conduits for transporting diamonds from the mantle to the surface.

The diamonds are carried to the surface in a mixture of magma, rocks, and minerals. Once this material cools, it forms kimberlite rock, which holds diamonds and can be mined.

5. Diamond Stability at the Surface

While diamonds are stable deep in the Earth's mantle, they remain stable on the surface as well because of the hardness and the strength of their atomic bonds. Diamonds that reach the surface in volcanic eruptions may remain buried for millions of years in kimberlite deposits, waiting to be discovered and mined.

6. Time Factor

The entire diamond formation process can take anywhere from **1 billion to 3.5 billion years**. Most of the diamonds we find today are ancient, having formed during the Earth's early history when our planet was much hotter and more geologically active than it is today.

7. Lab-Grown Diamonds (Alternative Method)

Though natural diamonds take eons to form, scientists have developed ways to mimic the natural diamond-making process in laboratories. These **lab-grown diamonds** are created using two primary methods:

- **High-Pressure High-Temperature (HPHT)**: Carbon is subjected to the same high-pressure, high-temperature conditions found in the Earth's mantle.
- **Chemical Vapor Deposition (CVD)**: Carbon-rich gases are used to deposit layers of carbon atoms

onto a substrate, building the diamond structure atom by atom.

While lab-grown diamonds have the same chemical and physical properties as natural diamonds, they can be produced in just weeks or months.

Conclusion

The formation of diamonds is a spectacular process that involves carbon being subjected to incredible heat and pressure deep within the Earth's mantle over billions of years. From these extreme conditions, one of nature's most sought-after and beautiful materials emerges. Brought to the surface by volcanic activity, diamonds then wait beneath the ground until they are discovered and transformed into the dazzling gemstones we admire today.

Whether formed naturally or in a lab, diamonds are a testament to the remarkable forces at work beneath our feet, creating beauty from the simplest of elements—carbon.

How is artificial intelligence trained?

Training artificial intelligence (AI) is a complex, multi-step process that involves feeding data into algorithms to enable the AI system to recognize patterns, make decisions, and improve over time. Here's a breakdown of how artificial intelligence is trained:

1. Defining the Task

The first step in training AI is clearly defining the task you want the AI to perform. This could range from image recognition (e.g., identifying objects in pictures) to natural language processing (e.g., understanding and generating human language), or more complex tasks like playing games, diagnosing diseases, or predicting stock market trends.

2. Choosing the Right Algorithm

Once the task is defined, the next step is selecting the appropriate algorithm. The algorithm is the mathematical model that learns from data. Common AI algorithms include:

- **Supervised Learning**: In supervised learning, the AI is trained on a labeled dataset, where the input

data is paired with the correct output. For example, if you're training an AI to recognize cats in images, you would provide the AI with thousands of images labeled as either "cat" or "not cat." The algorithm learns to map input data (the image) to the correct output (the label).
- **Unsupervised Learning**: In unsupervised learning, the AI is given data without explicit labels. It must find patterns or relationships in the data on its own. This is often used in tasks like clustering, where the AI groups similar data points together.
- **Reinforcement Learning**: Here, the AI learns by interacting with an environment and receiving feedback in the form of rewards or penalties. This method is commonly used in game playing and robotics. The AI is trained to maximize its reward over time by learning the best strategies through trial and error.
- **Deep Learning**: A subset of machine learning, deep learning uses neural networks with multiple layers to analyze data in increasingly abstract ways. This is often used in tasks like image recognition and language translation.

3. Data Collection and Preprocessing

Data is the fuel for AI. The more data you have, and the better its quality, the better your AI will perform. To train

Behind the Scenes

an AI, vast amounts of data are collected, often ranging from text, images, and videos to numerical data and sound recordings.

However, raw data is rarely perfect. It often contains noise, irrelevant information, or inconsistencies. Preprocessing the data is crucial to improving the quality of the training. This may include:

- **Cleaning the Data**: Removing any errors, duplicates, or outliers from the dataset.
- **Normalization**: Ensuring that the data is on a similar scale. For example, if training AI to predict house prices, it's important that square footage and number of rooms are scaled similarly so that one doesn't overpower the other.
- **Data Augmentation**: Generating new data from the existing dataset, such as flipping, rotating, or cropping images to increase the variety of the data without collecting new samples.

4. Training the Model

Once the data is ready, the algorithm is trained using this data. The model is initialized with random parameters (weights) and iteratively adjusts them to minimize errors.

- **Forward Propagation**: The input data passes through the model, layer by layer, until it produces an output (a prediction).
- **Loss Function**: The loss function measures how far the model's prediction is from the actual result. The loss (or error) tells the AI how well it's doing and guides the model in making adjustments to improve.
- **Backpropagation**: After the loss is calculated, the model uses backpropagation to adjust the weights of the neural network. This process repeats many times, with the model learning to minimize its loss and make more accurate predictions over time.
- **Gradient Descent**: This is the optimization algorithm that helps the AI model find the lowest possible error by adjusting its weights. The model follows the gradient (slope) of the loss function to find the minimum error.

This process is repeated through multiple **epochs** (passes through the entire dataset) until the AI model reaches an acceptable level of accuracy.

5. Validation and Tuning

After training, the model's performance is tested on a **validation set**—a portion of the data that the model hasn't seen during training. This is to ensure that the model can generalize well to new, unseen data rather than just memorizing the training examples.

During this step, the AI model may undergo **hyperparameter tuning**. Hyperparameters are external configurations that control the learning process, like the learning rate or batch size. Tuning these parameters can help improve the model's performance.

6. Testing the Model

Once the model is validated, it's tested on a **test set**, which is another set of unseen data. This gives a realistic measure of how the AI will perform in the real world. If the model performs well on this data, it can be considered ready for deployment.

7. Deployment and Monitoring

The final trained AI model is then deployed into a real-world environment. Whether it's a voice assistant, a recommendation system, or an autonomous vehicle, the model is put to use.

However, training doesn't stop here. The AI model must be continually monitored, as its performance may degrade over time due to changes in data (a phenomenon known as **data drift**). Regular retraining with updated data is essential to maintaining the accuracy and relevance of the AI system.

8. Transfer Learning

In many cases, instead of training a model from scratch, AI can be trained using **transfer learning**. Here, a pre-trained model (such as a neural network that's already learned from a vast dataset) is fine-tuned with new, specific data. This allows for faster training and improved accuracy, especially when dealing with smaller datasets.

9. Challenges in Training AI

Training AI is not without its challenges. Some common hurdles include:

- **Bias in Data**: If the training data is biased or unrepresentative, the AI model may make biased or unfair decisions. Ensuring diversity and fairness in the training data is crucial.
- **Overfitting**: This happens when the model learns the training data too well, including the noise or irrelevant details, resulting in poor performance on new data.
- **Data Privacy and Ethics**: When dealing with sensitive data, such as personal or medical information, data privacy and ethical considerations become paramount.

Conclusion

Training artificial intelligence is a highly iterative and data-intensive process. From choosing the right algorithm to collecting high-quality data, and from fine-tuning hyperparameters to testing in real-world environments, each step plays a vital role in the development of a reliable and effective AI model. As technology continues to advance, AI will only become more sophisticated, but at its core, AI training will always rely on human insight, creativity, and meticulous attention to detail.

Behind the Scenes

How Are Weather Forecasts Predicted?

Weather forecasting is a complex and fascinating process that involves the integration of science, technology, and data from a global network of monitoring systems. Despite how seamlessly weather predictions appear on our screens and smartphones, the science behind accurate forecasting is a blend of physics, atmospheric science, and advanced computing. Let's dive into the step-by-step process of how meteorologists predict the weather, transforming raw data into the reliable forecasts that help guide our daily lives.

1. Collecting Weather Data

The process of weather forecasting begins with the collection of vast amounts of meteorological data from around the world. Meteorologists gather this information from a range of sources, including:

- **Weather Stations:** These are ground-based stations that measure temperature, humidity, air pressure, wind speed, and direction. There are thousands of such stations strategically located across cities, rural areas, and remote regions.
- **Weather Satellites:** Orbiting high above the Earth, weather satellites continuously monitor the

atmosphere, oceans, and cloud formations. They capture images and data on storm systems, cloud cover, and the movement of weather patterns.
- **Weather Buoys:** Floating on the ocean's surface, buoys measure sea temperature, wave heights, and other oceanic conditions, which are crucial for understanding weather patterns that originate over the seas.
- **Radiosondes:** These are balloon-borne devices launched into the atmosphere that record data on temperature, humidity, and atmospheric pressure as they ascend to high altitudes. They provide a vertical profile of atmospheric conditions, essential for understanding the 3D structure of the atmosphere.
- **Radar Systems:** Doppler radar systems emit radio waves that bounce off raindrops, snowflakes, or hail, helping meteorologists see precipitation in real time and track severe weather systems like thunderstorms and hurricanes.

With this extensive network of data-collecting systems, meteorologists can observe and monitor the current state of the atmosphere across the globe.

2. Inputting Data into Weather Models

Once meteorologists gather this raw data, it's fed into highly sophisticated **numerical weather prediction**

(NWP) models. These computer models are simulations of the atmosphere that use mathematical equations to represent the physical laws governing weather, such as thermodynamics and fluid dynamics.

Popular weather models include:

- **Global Forecast System (GFS)** – Developed by the United States, it provides global weather forecasts up to 16 days in advance.
- **European Centre for Medium-Range Weather Forecasts (ECMWF)** – Known for its precision, this model is widely used by meteorologists around the world.
- **High-Resolution Rapid Refresh (HRRR)** – A short-range weather model that provides high-resolution forecasts for specific regions, ideal for predicting local weather patterns.

These models rely on supercomputers to handle the enormous volume of data and perform trillions of calculations per second. They divide the atmosphere into a 3D grid, with each cell representing a small region of the atmosphere. The model then calculates how the conditions in each cell—such as temperature, wind, and moisture—change over time based on physical equations.

3. Running Simulations and Creating Forecasts

With the data in place, the models simulate the behavior of the atmosphere over time, often running multiple times with slight variations in the starting conditions. These variations, known as **ensemble forecasting**, help predict a range of possible scenarios, taking into account the inherent uncertainties in initial data.

The result is a set of forecast outputs showing how temperature, pressure, wind, precipitation, and other factors are expected to evolve over hours, days, and even weeks. The models produce graphical outputs, maps, and numerical data, which meteorologists then analyze to identify trends and potential weather events.

4. Human Interpretation and Expertise

Even with advanced models, human expertise is crucial. Meteorologists interpret the model outputs, comparing results from multiple models to detect inconsistencies or areas where the models may have uncertainties. For example, certain weather models might excel at predicting large-scale systems like hurricanes but struggle with local phenomena like thunderstorms.

Meteorologists also consider historical weather patterns, local geography, and recent trends that models might overlook. This step is particularly important for making **short-term predictions** or when forecasting for regions with complex terrains, such as mountainous areas or coastal zones.

5. Generating and Communicating the Forecast

Once meteorologists have analyzed the model outputs, they generate weather forecasts that are shared with the public through various channels, such as television, radio, online platforms, and mobile apps. Forecasts are typically categorized into different time frames:

- **Short-range Forecasts:** Covering up to 72 hours. These are the most reliable and include detailed predictions like daily temperatures, wind speeds, and precipitation.
- **Medium-range Forecasts:** Extending from 3 to 7 days. These provide a broader view of weather trends, such as incoming cold fronts or storm systems.
- **Long-range Forecasts:** Covering 8 to 14 days and beyond. Long-range forecasts focus on general trends, such as shifts in weather patterns, rather than specific day-to-day events.

Meteorologists use visual aids like **weather maps**, animated models, and satellite images to help the public understand these forecasts. Communicating the level of certainty and potential impacts of weather events is crucial, especially when predicting severe weather conditions.

6. Continuous Monitoring and Updates

Weather is inherently dynamic, meaning conditions can change rapidly. To keep forecasts accurate, meteorologists continuously monitor new data from weather stations, satellites, and radar systems. As the atmosphere evolves, they update forecasts accordingly and issue warnings if necessary.

This ongoing process is particularly vital during extreme weather events, such as hurricanes, tornadoes, or severe thunderstorms, where updated forecasts can save lives by providing timely information to communities.

The Role of Advanced Technologies

Advancements in technology continue to improve the accuracy and reliability of weather forecasts. The use of **artificial intelligence (AI)** and **machine learning** allows meteorologists to detect patterns and refine predictions. Additionally, new satellite systems with higher resolution imagery provide more detailed observations, while increased computing power enables models to simulate weather with greater precision.

In the future, even more sophisticated models and data collection methods—like using drones to sample atmospheric conditions—will further enhance our ability to predict the weather with accuracy and detail that was once unimaginable.

Why Weather Forecasting Matters

Accurate weather forecasts play a critical role in modern society. They influence a wide range of activities, from agriculture and transportation to emergency management and public safety. Weather predictions allow farmers to optimize planting and harvesting schedules, help airlines avoid dangerous turbulence, and enable governments to issue early warnings before severe weather strikes.

With climate change contributing to more unpredictable and extreme weather patterns, the importance of reliable weather forecasting has never been greater. Meteorologists continue to push the boundaries of what's possible, ensuring that we stay informed and prepared, no matter what nature throws our way.

By understanding the science and technology behind weather forecasting, we gain a deeper appreciation for the dedicated efforts of meteorologists who work tirelessly to keep us one step ahead of the elements.

Behind the Scenes

How is 3D Printing Done?

3D printing, also known as **additive manufacturing**, is a revolutionary process that creates three-dimensional objects by layering materials based on digital designs. Unlike traditional manufacturing, which often involves subtracting material (cutting, drilling, etc.), 3D printing adds material layer by layer to build an object from the ground up. This process is used across various industries, from medical devices and automotive parts to fashion and even food. Here's a detailed look into how 3D printing is done:

---

1. Creating a Digital Design

The first step in 3D printing is creating a **digital 3D model** of the object to be printed. This can be done in several ways:

- **Computer-Aided Design (CAD) Software**: Designers use CAD software (such as AutoCAD, Blender, or SolidWorks) to develop detailed, precise models.
- **3D Scanning**: If an existing object needs to be replicated, a 3D scanner captures the object's dimensions and creates a digital model.

Behind the Scenes

- **Downloading Pre-Existing Models**: Numerous online repositories, like Thingiverse, offer ready-made 3D models for printing.

Once the design is finalized, it's converted into an STL file (StereoLithography file) or a similar format that a 3D printer can read.

---

2. Slicing the Design

After creating the 3D model, the next step is to **slice** it into thin horizontal layers using slicing software (e.g., Cura or Simplify3D). Slicing converts the 3D model into instructions (G-code) for the printer to follow. These instructions include:

- **Layer height**: Each layer's thickness, which can range from fractions of a millimeter to several millimeters.
- **Infill density**: The amount of material inside the object, often expressed as a percentage. A higher percentage creates a more solid structure.
- **Print speed and temperature**: Instructions on how fast the printer should move and at what temperature to melt the material.

The sliced file is then ready to be sent to the 3D printer.

3. Setting Up the 3D Printer

Before printing begins, the printer must be prepared:

- **Material Loading**: Depending on the type of 3D printer, different materials are used. The most common materials are:
    - **Plastic Filaments (FDM printers)**: PLA and ABS are the most popular.
    - **Resin (SLA printers)**: Liquid photopolymer that hardens when exposed to light.
    - **Powder (SLS printers)**: Fine powders like nylon or metal that are fused together.
- **Printer Calibration**: Ensuring that the printer's build platform is level and the printhead is properly aligned is crucial for a successful print.

4. The Printing Process

Once everything is set, the actual printing begins. Depending on the type of 3D printer used, the process varies slightly:

- **Fused Deposition Modeling (FDM)**:

Behind the Scenes

- - The most common type of 3D printing, FDM works by extruding melted thermoplastic filament through a heated nozzle.
  - The printer moves layer by layer, depositing thin lines of material according to the sliced model.
  - Each layer cools and solidifies before the next one is added, gradually building the object from the bottom up.
- **Stereolithography (SLA):**
  - In SLA printing, a vat of liquid resin is used. A laser beam traces the pattern of the object's layer onto the surface of the resin.
  - The light-sensitive resin hardens wherever the laser hits it, solidifying one layer at a time.
  - The platform lowers slightly between layers, allowing the next one to be traced.
- **Selective Laser Sintering (SLS):**
  - SLS printers use powdered materials (such as nylon, ceramic, or metal). A high-powered laser fuses the particles together layer by layer.
  - After each layer is sintered, a new layer of powder is spread across the build platform for the next layer to be printed.

The process continues until the object is fully formed. Depending on the complexity and size of the object, printing can take anywhere from a few minutes to several hours or even days.

5. Post-Processing

Once the object is printed, it usually requires some level of post-processing to achieve the final desired result:

- **Removing Supports**: Many 3D printed objects are built with support structures to prevent sagging during printing. These are carefully removed afterward.
- **Sanding and Smoothing**: For a smoother finish, the object may need to be sanded or polished, especially if layer lines are visible.
- **Curing (for SLA prints)**: SLA prints are often placed under UV light to fully cure and harden the resin.
- **Painting and Finishing**: Depending on the use, the object might be painted, dyed, or coated for extra durability or aesthetic appeal.

6. Applications of 3D Printing

The versatility of 3D printing allows it to be used in a wide range of industries, including:

- **Healthcare**: 3D printers are used to create prosthetics, dental implants, and even bio-printed organs and tissues.
- **Aerospace and Automotive**: Complex parts for aircraft and cars can be rapidly prototyped or manufactured using 3D printing.
- **Education and Art**: 3D printing enables students and artists to bring their concepts to life, from architectural models to sculptures.
- **Fashion and Jewelry**: Designers use 3D printing to create intricate, custom pieces.
- **Food**: Specialized 3D printers can layer edible materials like chocolate or dough to create custom foods.

Conclusion

3D printing is transforming industries by providing a faster, more customizable way to manufacture objects. From prototyping to creating finished products, the technology has proven to be a game-changer in how we make things. As it continues to evolve, 3D printing is poised to become even more accessible, enabling individuals and businesses alike to bring their ideas to life with precision and creativity.

Behind the Scenes

How are solar panels made?

Solar panels are a marvel of modern engineering, transforming sunlight into usable electricity through a fascinating process. Here's an in-depth look at how they're made, from raw materials to the final product:

1. Raw Materials: Silicon Extraction

The primary material used to make solar panels is **silicon**, one of the most abundant elements on Earth. Silicon is sourced from **silica**, a compound found in sand, quartz, and other minerals. The process begins by mining quartz-rich sand, which is then refined into pure silicon through a series of steps that involve extreme heat (up to 1800°C) to remove impurities. This highly pure silicon is crucial, as impurities can affect the efficiency of solar panels.

2. Ingot Formation: Creating Silicon Blocks

Once pure silicon is obtained, the next step is to create **silicon ingots**. The process used is called the **Czochralski Process**, where silicon is melted in a furnace, and a small piece of crystalline silicon (known as a seed) is dipped into the molten silicon. As the seed is slowly pulled out, the molten silicon crystallizes around it, forming a large cylindrical block of pure silicon called an ingot. These ingots can be either **monocrystalline** (a single continuous crystal) or **polycrystalline** (composed of many small crystals).

- **Monocrystalline solar panels** are more efficient because the silicon crystals are perfectly aligned, allowing for better electron flow. However, they are more expensive to produce.
- **Polycrystalline panels** are less expensive but slightly less efficient due to the multiple crystal structures.

3. Wafer Production: Slicing the Silicon Ingots

The silicon ingots are then sliced into thin, flat **wafers** using a high-precision wire saw. These wafers, usually just a few millimeters thick, will serve as the base for the solar cells. The thinner the wafers, the less silicon is wasted, which helps reduce manufacturing costs.

4. Doping: Creating a Semiconductor Layer

To turn the silicon wafers into functional solar cells, the wafers need to be treated so that they can generate electricity when exposed to sunlight. This is done through a process called **doping**, where a small amount of another element is added to the silicon to alter its electrical properties.

Behind the Scenes

- The silicon wafer is coated with **phosphorus** on the top, creating a **negative charge layer** (N-type silicon).
- The bottom of the wafer is treated with **boron**, which introduces a **positive charge layer** (P-type silicon).

This creates a **P-N junction**, which is crucial for the photovoltaic effect, the process that converts sunlight into electricity. When sunlight hits the solar cells, it excites the electrons in the silicon, creating an electric current.

5. Anti-Reflective Coating: Enhancing Efficiency

Silicon naturally reflects a lot of sunlight, which can reduce the efficiency of the solar cells. To combat this, the wafers are coated with an **anti-reflective layer**. This layer, usually made from silicon nitride, helps reduce the amount of light that is reflected, ensuring that more sunlight is absorbed by the cell. This improves the overall efficiency of the solar panel.

6. Metal Contacts: Collecting the Electric Current

Once the silicon wafers are treated, **metallic conductors** are added to the surface of the wafers. These metal strips, often made of **silver**, form a grid-like pattern on the top and a solid metal layer on the back of the wafer. The purpose of these contacts is to collect the electric current generated by the excited electrons when sunlight strikes the solar cells.

The front grid collects the electrons that flow toward the surface, while the back contact gathers the electrons that flow toward the bottom of the cell, allowing electricity to flow through the circuit.

7. Assembling the Solar Cells: Creating Solar Panels

The individual solar cells are fragile, so they are carefully handled and assembled into larger units to create a solar panel. These cells are connected in series and parallel circuits to achieve the desired voltage and current levels.

The solar cells are placed between two protective layers of **EVA (ethylene-vinyl acetate)**, which help cushion the cells and protect them from moisture. A **tempered glass** layer is added to the front of the panel to protect the cells from environmental factors like hail, wind, and rain. The back is typically made from a durable plastic or polymer sheet.

8. Framing: Adding Durability

The assembled solar panel, now encased in glass and a protective backsheet, is placed into a **metal frame** made of aluminum. The frame provides structural support and protects the panel from damage during transport and

Behind the Scenes

installation. It also includes mounting holes and grooves for easy attachment to rooftops or solar farms.

9. Testing and Quality Control

Before the solar panels are ready to be shipped, they undergo a series of **quality control tests** to ensure they perform as expected. These tests check the panel's ability to generate electricity under different lighting conditions, its resistance to environmental stressors (such as heat, cold, and moisture), and its overall durability. Panels that pass these tests are deemed ready for use and are labeled with a **power rating**, indicating the amount of electricity they can produce under standard conditions (measured in watts).

10. Installation and Operation

Once manufactured, solar panels are ready to be installed in homes, businesses, or large solar farms. During installation, multiple panels are connected together to form a **solar array**, which is wired to an **inverter** that converts the direct current (DC) electricity generated by the panels into alternating current (AC) electricity, which can be used by standard appliances and sent back to the grid.

The Final Product: Solar Energy in Action

Behind the Scenes

Solar panels, after this precise and meticulous production process, are capable of harnessing the power of the sun to generate clean, renewable energy. Over the past few decades, the efficiency of solar panels has significantly improved, while production costs have dropped, making solar power a viable alternative to fossil fuels for many people around the world.

By converting sunlight into electricity with no pollution, greenhouse gas emissions, or fuel costs, solar panels play a critical role in **sustainable energy production** and are a key technology in the global effort to combat climate change.

Behind the Scenes

How is Cheese Produced?

Cheese, one of the world's most beloved foods, has been produced for thousands of years, with a variety of types that differ in texture, flavor, and appearance. The process of cheese production involves a delicate balance of science and craftsmanship, taking milk and transforming it into a wide array of cheeses, each with its own unique characteristics. This fascinating transformation is rooted in chemistry, biology, and a deep understanding of traditional techniques. Let's dive into how cheese is produced.

Step 1: Milk Collection and Preparation

The cheese-making process begins with fresh milk, which can come from cows, goats, sheep, or even buffalo. The quality and type of milk play a critical role in determining the flavor, texture, and type of cheese produced. Milk used for cheese production is carefully monitored for quality, as factors such as fat content, protein levels, and freshness are vital to the success of the process.

Before cheese can be made, the milk is usually **pasteurized** to eliminate harmful bacteria. However, some cheeses, especially traditional varieties, are made from raw milk to retain natural flavors and bacteria that contribute to the cheese's distinct taste. Once pasteurized, the milk is cooled and prepared for the next stage of the process.

Step 2: Coagulation

Coagulation is the process by which milk is transformed from a liquid to a semi-solid state. To do this, cheesemakers introduce a **starter culture** into the milk. This culture consists of beneficial bacteria that begin fermenting the lactose (milk sugar), turning it into lactic acid. This acidification of the milk is key to cheese production, as it helps create the right environment for the formation of curds.

Next, **rennet** is added to the milk. Rennet is an enzyme traditionally sourced from the stomach lining of young calves, though modern cheese production also uses vegetarian or microbial alternatives. The rennet causes the milk proteins, primarily casein, to coagulate and form curds. The curds are the solid part of the milk, while the remaining liquid is called whey.

Step 3: Cutting the Curds

Once the curds have formed, they are cut into smaller pieces using a tool called a curd knife. The size of the curds can vary depending on the type of cheese being made. For example, smaller curds are typically used for harder cheeses like Parmesan, while larger curds are used for softer varieties like Brie.

Behind the Scenes

Cutting the curds serves a dual purpose: it releases more whey, which must be drained, and it allows for greater control over the cheese's final texture. As the curds are cut, they begin to shrink, firm up, and expel additional whey.

Step 4: Cooking and Stirring

After the curds are cut, they are gently heated and stirred to help release even more whey. This step is known as **cooking the curds**, though the temperature and duration of cooking depend on the type of cheese being produced. For instance, high-temperature cooking creates firmer curds for hard cheeses, while lower temperatures result in softer curds for cheeses like Mozzarella.

As the curds cook, they continue to lose moisture and whey, which is essential for the development of the cheese's texture and consistency. Cheesemakers closely monitor this step to ensure the curds reach the desired firmness.

Step 5: Draining the Whey

Once the curds have been cooked and reached the desired texture, the whey is drained off. This can be done by simply pouring off the whey or using special molds that allow the whey to drain naturally over time. At this stage, the curds are now solidifying and beginning to take shape.

Behind the Scenes

In many cases, the curds are placed into cheese molds, which further help drain the whey and shape the cheese into its final form. These molds can be round, square, or any number of shapes depending on the cheese variety.

Step 6: Salting

Salt plays an essential role in cheese production, acting as a preservative, enhancing flavor, and helping to draw out moisture. There are two primary methods of salting cheese:

- **Dry Salting:** Cheese curds are sprinkled with salt after being placed in molds.
- **Brining:** The formed cheese is soaked in a saltwater solution, or brine, which allows the salt to penetrate the cheese evenly.

Salt also helps form the outer rind of some cheeses and contributes to the final flavor and texture. Different cheeses have different levels of salt, depending on their variety and region of origin.

Step 7: Pressing (Optional)

For harder cheeses, the curds are often pressed to remove any remaining whey and to form a more compact, solid texture. The curds are placed in molds and subjected to pressure, which compacts them and helps create the dense, firm texture associated with cheeses like Cheddar or Gouda.

The pressing time can range from a few hours to several days, depending on the desired firmness of the cheese. Softer cheeses, such as Brie or Camembert, typically skip the pressing step and retain a looser, creamier texture.

Step 8: Aging (Affinage)

Not all cheeses require aging, but for those that do, this final step is where the true magic happens. **Aging** or **affinage** is the process of letting the cheese mature over time in controlled environments with specific temperature and humidity levels.

During aging, the cheese develops its full flavor profile, texture, and aroma. The duration of aging varies widely— some cheeses age for just a few weeks, while others may take years to reach maturity. For example:

- **Fresh cheeses** like Ricotta or Feta are consumed without aging.
- **Soft-ripened cheeses** like Brie may age for a few weeks.
- **Hard cheeses** like Parmesan can age for up to several years, developing intense, complex flavors.

Throughout the aging process, natural molds, bacteria, and enzymes continue to interact with the cheese, developing its rind, texture, and taste.

Step 9: Packaging and Distribution

Once the cheese has reached its desired age and flavor, it's ready to be packaged and distributed. Cheeses are often wrapped in wax paper, plastic, or special breathable materials that allow them to continue aging slightly while protecting them from spoilage.

Some cheeses, especially artisanal varieties, are sold directly from the aging rooms, while others are carefully packaged for transport to markets around the world.

Conclusion

Cheese production is both an art and a science, with centuries of tradition and innovation behind every wheel, wedge, or slice. From the initial collection of milk to the final bite on your plate, the journey of cheese is a fascinating process that combines nature's raw ingredients with human expertise and creativity. Whether it's the fresh tang of a goat cheese or the sharp bite of an aged Cheddar, every cheese carries within it the story of how it was made, crafted to delight taste buds across the globe.

How Are Movies Made from Script to Screen?

Creating a movie is a complex, multi-stage process that involves a blend of creativity, technical expertise, and meticulous planning. From the initial spark of an idea to the final version projected on the big screen, filmmaking is a fascinating journey that brings together a diverse team of professionals, each playing a vital role in transforming words on a page into a visual story. Let's take a look at the key stages involved in making a movie from script to screen.

1. The Idea and Scriptwriting

Every film begins with an idea. Whether it's a concept for an original story, an adaptation of a book, or a real-life event, this initial spark is the foundation of the film. The screenwriter takes this idea and crafts it into a screenplay, which serves as the blueprint for the movie. The screenplay includes dialogue, character descriptions, and detailed scene-by-scene action.

A screenplay often goes through several drafts, with input from producers, directors, and sometimes even actors, before it's finalized. At this stage, the goal is to make the script compelling, visual, and suitable for production.

Behind the Scenes

2. Development and Pre-production

Once the script is finalized, the development phase begins. Producers pitch the script to studios or investors to secure funding. This process can take years, depending on the scope of the film and its budget. Once funding is secured, the project moves into pre-production, where the groundwork is laid for shooting.

Key activities during pre-production include:

- **Casting**: The director and casting director select actors for the roles, with major stars sometimes being attached early on to attract financing.
- **Location Scouting**: The team searches for locations that fit the script's settings or decides if certain scenes will be filmed on constructed sets.
- **Storyboarding**: Visual representations of scenes are created, mapping out how each shot will look and how it fits into the overall narrative.
- **Budgeting and Scheduling**: The film's budget is finalized, and shooting schedules are created to ensure everything runs smoothly during production.

3. Production

The production phase, or principal photography, is when the actual filming takes place. This is the most visible part of the filmmaking process, where actors perform in front of the camera, and the director leads the crew to capture the script's action.

Key elements of production include:

- **Directing**: The director oversees all creative aspects, guiding actors, deciding camera angles, and ensuring that the visual storytelling aligns with the script's vision.
- **Cinematography**: The director of photography (DP) is responsible for lighting, camera work, and the overall look of the film. They work closely with the director to capture the desired aesthetic.
- **Set Design and Costume**: The production designer and costume designer ensure the film's sets and costumes are authentic to the story and period, contributing to the world-building.
- **Sound and Music**: During filming, sound technicians capture dialogue and ambient sounds. Some of this may be re-recorded later in post-production to achieve perfect quality.

4. Post-production

Once filming wraps, the project moves into post-production, where the raw footage is transformed into a polished movie. This phase can be just as time-consuming and crucial as production itself, involving a team of editors, sound designers, visual effects artists, and musicians.

Key post-production activities include:

- **Editing**: The editor pieces together the film, cutting out unwanted scenes and arranging the footage to ensure a coherent and engaging story. Editors work closely with the director to match the film's tone and pacing to their vision.
- **Sound Design**: Foley artists create sound effects, such as footsteps or creaking doors, to enhance the auditory experience. Dialogue is cleaned up, and sound mixing balances all audio elements.
- **Visual Effects (VFX)**: If the movie requires computer-generated imagery (CGI) or special effects, the visual effects team works on these elements, adding them to the film.
- **Music Scoring**: The composer creates the film's soundtrack, which is then recorded and integrated into the final cut. Music plays a significant role in enhancing the emotional impact of scenes.
- **Color Grading**: The footage is color-corrected to achieve the desired mood and look, ensuring consistency throughout the film.

5. Distribution and Marketing

Behind the Scenes

With post-production completed, the film is now ready for distribution. This process involves finding the best way to bring the movie to audiences, whether through theatrical release, streaming platforms, or film festivals.

- **Film Festivals**: Many filmmakers debut their work at festivals to generate buzz and attract distributors.
- **Theatrical Release**: If a film secures a distributor, it will be shown in theaters. The release strategy (wide or limited release) depends on the film's appeal and potential audience size.
- **Marketing**: Trailers, posters, press interviews, and social media campaigns help create awareness and excitement for the film's release. A strong marketing campaign can determine the movie's success at the box office.

Conclusion

From the initial concept to the glitzy premiere, the process of making a movie is a remarkable blend of artistry, technical skill, and collaboration. Every movie represents the work of hundreds or even thousands of people, each contributing their expertise to bring a story to life. Whether it's a small indie film or a blockbuster with a massive budget, the journey from script to screen is an intricate dance that demands passion, precision, and creativity at every step.

Behind the Scenes

Next time you sit down to watch a movie, remember the incredible effort that goes into every scene. Behind the camera, a world of craft and magic is at work, making the impossible possible and turning imagination into reality.

Behind the Scenes

How Are Musical Instruments Made?

Musical instruments, the tools that bring melody and rhythm to life, have been a cornerstone of human expression for thousands of years. From the primitive flutes made of bone by ancient civilizations to the highly engineered electric guitars of today, musical instruments come in many shapes, sizes, and sounds. But how are these intricate instruments crafted? The process behind making musical instruments is a beautiful blend of science, art, and craftsmanship. Let's take a look at how some of the most popular musical instruments are made.

String Instruments (Violins, Guitars, Cellos)

1. Selecting the Right Wood: The foundation of most string instruments is the wood. Instrument makers, also known as luthiers, carefully select different types of wood based on their sound properties. For example, violins are often made from spruce for the top (soundboard) and maple for the back and sides. Each type of wood affects the resonance and tonal quality of the instrument.

2. Shaping and Carving: Once the wood is selected, the shaping process begins. For violins and cellos, the wood is carved and shaped by hand to precise dimensions to ensure the right sound quality. The top and back are meticulously carved into arched shapes, which help amplify sound. The neck, scroll, and fingerboard are crafted with equal attention to detail, allowing for ease of play and proper sound projection.

3. Assembling the Instrument: Once the individual components are shaped, they are carefully glued together. Unlike modern adhesives, traditional instruments often use hide glue, which has acoustic benefits and can be undone if repairs are needed. The body of the instrument is assembled in stages to ensure structural stability and proper alignment.

4. Varnishing and Finishing: To protect the wood and enhance its appearance, the instrument is coated with varnish. This step is essential because the varnish not only adds aesthetic appeal but also influences the sound. Luthiers apply varnish in thin layers, often over several days or weeks, to ensure the instrument retains its acoustic properties.

5. Stringing and Fine-Tuning: The final step is adding the strings and fine-tuning the instrument. This process involves installing the bridge, sound post, and strings, followed by precise tuning to ensure the instrument is ready to produce music. For string instruments like violins and guitars, the choice of string materials—nylon, gut, or metal—can greatly affect the sound.

Wind Instruments (Flutes, Clarinets, Trumpets)

1. Materials Selection: Wind instruments are often made from metals (brass, silver) or wood (for woodwinds like clarinets and oboes). The choice of material plays a key role in determining the tone and timbre of the instrument. For example, brass instruments like trumpets are made from brass alloys, while flutes can be made from silver or nickel.

2. Shaping the Body: For instruments like flutes and trumpets, the body starts as a flat sheet of metal that is rolled into a cylindrical shape. The tube is then welded or soldered together to form the main body. For clarinets and oboes, the wood is hollowed out using specialized tools to create the body, with precise attention to internal dimensions.

3. Crafting Keys and Mechanisms: Wind instruments rely on intricate key systems to control the pitch. These keys are carefully machined and attached to the body with screws, rods, and springs. Each key is designed to seal holes on the instrument's body, altering the length of the air column when pressed, which changes the pitch.

4. Tuning and Adjusting: Once the body and key mechanisms are in place, the instrument is fine-tuned. Craftspeople make micro-adjustments to the bore (the internal tube) to ensure that the instrument plays in tune across all registers. This step is crucial for ensuring that wind instruments can produce a clear, accurate tone.

Percussion Instruments (Drums, Marimbas)

1. Frame and Shell Construction: For percussion instruments like drums, the body or shell is typically made from wood or metal. Wooden drums are usually crafted by layering thin strips of wood into a circular mold, while metal drums are shaped from steel or brass. Marimbas and xylophones, on the other hand, are made from hardwood bars carefully carved to resonate at specific frequencies.

2. Skinning the Drums: The drum's head, which produces the sound, is traditionally made from animal hide but can also be made from synthetic materials like Mylar. The drumhead is stretched over the shell and secured with a tension rod system that allows for tuning. The tighter the drumhead is stretched, the higher the pitch it will produce.

3. Tuning and Adjustments: Drums and other percussion instruments require careful tuning to ensure that they produce the desired pitch. In the case of instruments like timpani, which are used in orchestras, this tuning is

Behind the Scenes

extremely precise, allowing the drum to play specific notes.

Electronic Instruments (Synthesizers, Electric Guitars)

1. Designing the Circuitry: For electronic instruments like synthesizers, the heart of the instrument is the circuitry. Engineers design circuit boards that can generate, modify, and amplify sound waves. Synthesizers often use oscillators, filters, and modulators to shape the sound. The design process involves careful planning to ensure the instrument can produce a wide range of sounds.

2. Building the Body: For electric guitars, the body is typically made from solid wood such as alder, mahogany, or ash. The body is then shaped and carved to create the desired look and feel. The neck is often bolted or glued to the body, and the fretboard is carefully positioned to ensure accurate intonation.

3. Installing Pickups and Electronics: The electronics in electric guitars include pickups, which capture the vibration of the strings and convert them into electrical signals. These pickups are installed into the guitar body and connected to tone and volume controls, allowing the musician to shape their sound. Synthesizers, on the other hand, use digital or analog components to generate electronic sound waves that can be manipulated with knobs and sliders.

4. Final Calibration and Testing: Once the instrument is assembled, it goes through a final calibration process. For electric guitars, this includes adjusting the action (string height), intonation, and tuning. Synthesizers undergo sound tests to ensure the electronics are functioning properly and that the instrument can produce the full range of tones expected.

The Craftsmanship Behind the Music

The creation of musical instruments is a perfect marriage of art and science. Whether handcrafted by artisans or mass-produced with precision machinery, every instrument is a product of careful design, meticulous attention to detail, and a deep understanding of how sound works.

From the selection of materials to the final tuning, each step in the process impacts the quality and character of the instrument. And while modern technology has revolutionized the way instruments are made, traditional craftsmanship still plays a vital role in bringing musical instruments to life. So, the next time you pick up a guitar, sit down at a piano, or listen to the sweet sounds of a flute, you'll know the incredible journey each instrument has undergone before it reaches your hands.

Behind the Scenes

How Is Space Debris Cleaned Up?

Space debris, also known as space junk, is a growing problem orbiting our planet. Consisting of defunct satellites, spent rocket stages, fragments from collisions, and even tiny bits of paint, this debris poses a serious risk to both manned and unmanned spacecraft. With more satellites being launched into space each year, the amount of debris continues to increase, leading to heightened concerns about potential accidents that could affect vital space missions and infrastructure.

Cleaning up space debris is no small feat. The challenge stems from the fact that this junk is orbiting Earth at incredible speeds—up to 28,000 kilometers per hour—making it difficult to capture and remove. However, innovative solutions are being developed to tackle this problem head-on. Let's explore some of the most advanced and promising techniques being used to clean up space debris.

1. Laser Ablation

One of the most futuristic methods for cleaning space debris involves the use of lasers. Scientists have developed a process called **laser ablation**, where high-powered ground-based or space-based lasers are fired at small pieces of debris. The laser doesn't destroy the debris; rather, it heats up the surface of the object, causing it to vaporize a tiny bit of material. This vaporization creates a small thrust, pushing the debris into a lower orbit, where it eventually burns up as it re-enters Earth's atmosphere.

Laser ablation is still in its experimental stages, but it holds significant promise as a method for removing smaller pieces of debris, which are among the hardest to track and the most dangerous to spacecraft.

2. Harpoons and Nets

A more hands-on approach to debris removal involves the use of **harpoons** and **nets**. This technique is designed to capture larger pieces of debris, such as defunct satellites or old rocket parts. Space agencies like the European Space Agency (ESA) have been experimenting with these methods in their cleanup missions.

For instance, ESA's mission, **RemoveDEBRIS**, tested the use of a harpoon to physically spear a piece of space junk and reel it back into a containment system. Similarly, the net system was deployed to ensnare larger objects. Once captured, the debris can either be directed toward Earth's atmosphere to burn up upon re-entry or placed in a safer, higher orbit away from active satellites.

Behind the Scenes

3. Robotic Arms and Tugging Satellites

In some cases, more delicate handling is needed. This is where **robotic arms** come into play. Some satellites are being equipped with robotic arms that can grab onto larger pieces of debris. These robotic arms then "tug" the debris into a new, safer orbit, or guide it toward Earth's atmosphere for destruction.

The **ClearSpace-1** mission by ESA is scheduled to be one of the first major attempts to use a robotic arm for space debris removal. The spacecraft will use its robotic arm to capture a defunct satellite, bringing it back toward Earth to safely burn up upon re-entry.

4. Electrodynamic Tethers

Another cutting-edge technology for space debris cleanup is the use of **electrodynamic tethers**. These long, conductive cables are attached to spacecraft and generate electricity as they move through Earth's magnetic field. By deploying these tethers and attaching them to pieces of debris, the spacecraft can create drag that slows down the debris. As the debris slows, it loses altitude and eventually falls back into Earth's atmosphere, where it burns up.

Electrodynamic tethers are particularly useful for removing smaller debris and objects that have been in orbit for many years.

5. Drag Sails

For satellites that have reached the end of their operational life, one potential solution is the use of **drag sails**. These are large, lightweight structures that can be deployed from the satellite itself. The sail increases the surface area of the satellite, creating more drag from the thin atmosphere present even in low Earth orbit.

The increased drag causes the satellite to lose altitude more quickly, speeding up its re-entry into the atmosphere, where it can safely burn up. This solution is designed to prevent future space junk from accumulating by providing satellites with a self-cleaning mechanism.

6. Space Sweepers

Several private companies and space agencies are working on spacecraft known as **space sweepers**—vehicles specifically designed to collect and remove debris from orbit. These spacecraft are equipped with various tools like nets, harpoons, and robotic arms to capture different types of debris.

One notable project is **Astroscale**, a private company that has been developing a system to capture dead satellites. Astroscale's **ELSA-d** mission successfully demonstrated the technology's ability to capture and deorbit objects in low Earth orbit, marking an important milestone in space debris removal efforts.

7. International Cooperation and Prevention

Beyond removing existing debris, preventing new debris from accumulating is critical to solving the problem long-term. International space agencies and organizations like the **United Nations Office for Outer Space Affairs (UNOOSA)** have implemented guidelines aimed at reducing the creation of space junk. These include regulations that require satellite operators to deorbit their spacecraft after their missions end or to place them in "graveyard orbits" where they won't interfere with operational satellites.

Additionally, improving satellite design by incorporating technologies like onboard propulsion systems and drag sails can help minimize future debris.

The Future of Space Cleanup

While no single solution is likely to solve the problem of space debris, the combined efforts of laser technology, harpoons, nets, robotic arms, and international regulations provide a promising way forward. The cleanup of space debris is an ongoing challenge, but with continued innovation and global cooperation, we can ensure that space remains a safe and sustainable environment for exploration, communication, and research.

As humanity pushes further into the final frontier, the importance of cleaning up our cosmic backyard cannot be understated. From protecting valuable satellites to ensuring the safety of future crewed space missions, space debris removal is an essential part of space exploration in the 21st century.

Behind the Scenes

How is DNA Testing Done?

DNA testing, also known as genetic testing, is a powerful scientific tool that allows us to analyze an individual's genetic code to uncover information about ancestry, health, relationships, and even identity. Since its inception, DNA testing has revolutionized fields such as medicine, forensic science, and genealogy. But how exactly is this intricate process done? Let's dive into the science behind DNA testing and explore the steps involved in unraveling the unique blueprint that makes you, you.

The Basics of DNA

DNA, or deoxyribonucleic acid, is the molecular blueprint of life. It contains the genetic instructions that guide the development, functioning, and reproduction of all living organisms. In humans, DNA is organized into chromosomes and is found in almost every cell of the body. This genetic material is composed of four chemical bases—adenine (A), thymine (T), cytosine (C), and guanine (G)—which form a unique sequence in each individual. It is this sequence that genetic testing analyzes to provide insights into an individual's characteristics, traits, or lineage.

Step 1: Collecting a DNA Sample

Behind the Scenes

The first step in DNA testing is **sample collection**. DNA can be collected from a variety of biological materials such as saliva, blood, hair follicles, skin cells, or even bone. The most common and non-invasive method for modern DNA testing is a **saliva or cheek swab**. Using a sterile cotton swab, cells are collected from the inside of the cheek. This method is easy, painless, and can be done at home or in a lab setting.

Step 2: Extracting the DNA

Once the sample is collected, the next step is to **extract the DNA** from the cells. In a lab, technicians use a chemical process to break down the cell membrane and release the DNA. This process typically involves using enzymes and chemical solutions that break apart the proteins and fats surrounding the DNA while keeping the genetic material intact.

Once the DNA is freed from the cells, it is separated from other cellular debris using a centrifuge. The result is a purified DNA sample that is ready for analysis.

Step 3: Amplifying the DNA Using PCR

After extraction, the amount of DNA in a typical sample is too small to analyze directly. Therefore, scientists use a process known as **Polymerase Chain Reaction (PCR)** to amplify the DNA. PCR is a technique that allows specific

Behind the Scenes

segments of DNA to be copied millions or billions of times in just a few hours. This ensures that there is enough DNA to work with for further testing.

PCR works by heating the DNA to separate the two strands, then cooling it so that a primer (a short piece of synthetic DNA) can attach to the target sequence. An enzyme called **Taq polymerase** then builds a new strand of DNA by adding the correct nucleotides, duplicating the segment over and over. This process is repeated in cycles to exponentially amplify the DNA.

Step 4: Analyzing the DNA

With sufficient DNA available, the next step is to analyze the specific regions of the genetic code. Different types of DNA tests are designed to look for specific markers, mutations, or sequences, depending on the purpose of the test.

- **Ancestry Testing:** For genealogical purposes, DNA testing looks at specific markers known as **single nucleotide polymorphisms (SNPs)**. These markers can reveal genetic similarities between individuals or populations and help trace ancestry and ethnicity.
- **Paternity Testing:** For paternity or relationship testing, the DNA is compared between two individuals (typically the child and potential father) at specific locations called **Short Tandem Repeats**

(STRs). If a significant number of these markers match, the test can determine biological relationships.
- **Medical Testing:** Medical DNA tests look for specific **genetic mutations** associated with hereditary diseases or conditions. For example, tests may check for mutations in the **BRCA1 or BRCA2** genes, which are linked to an increased risk of breast and ovarian cancer.

The genetic data is analyzed using sophisticated laboratory equipment, such as **sequencers**, which read and record the DNA sequence. Modern sequencing technologies are incredibly fast and accurate, allowing labs to process large amounts of genetic data in a short period.

Step 5: Interpreting the Results

Once the analysis is complete, the raw genetic data needs to be interpreted. Scientists compare the tested DNA sequences with known reference genomes or genetic markers. For medical or paternity testing, the results are straightforward: either a particular genetic mutation is present or it's not, or a match in paternity testing is either confirmed or ruled out.

For ancestry testing, the results are often more complex. DNA testing companies typically have large databases of genetic information that allow them to compare your DNA

Behind the Scenes

with reference populations. Based on this comparison, they provide estimates of your ethnic composition or ancestral origin. These results are usually presented in easy-to-understand percentages, maps, or family tree forms.

Step 6: Reporting the Results

After the interpretation, the results are compiled into a report and provided to the person who submitted the sample. Depending on the type of test, the report may include:

- **Ancestry breakdown** (percentages of ethnicity, migration patterns, etc.)
- **Health risks** or predispositions based on genetic markers (for medical testing)
- **Paternity or familial relationship results**
- **Carrier status** for genetic conditions

Some testing services also offer personalized recommendations for lifestyle, diet, or medical care based on the genetic data, although this varies depending on the test.

DNA Testing in Forensic Science

DNA testing also plays a pivotal role in forensic science, where it is used to solve crimes and identify victims. In criminal investigations, DNA from a crime scene is compared to a suspect's DNA to establish a match. Even small amounts of DNA—like a single hair or skin cell—can be enough to identify a person.

DNA testing has led to the exoneration of wrongfully convicted individuals and has been critical in cold case investigations, where samples collected years or even decades ago are now tested with modern technology.

The Future of DNA Testing

The future of DNA testing is bright, with advancements in technology making it faster, cheaper, and more accessible. Whole-genome sequencing, which reads the entirety of an individual's DNA, is becoming more common and is expected to unlock even more insights into human health, behavior, and ancestry.

Furthermore, researchers are continually discovering new applications for DNA testing, from predicting disease risk to developing personalized medical treatments. As science continues to unlock the secrets of our genetic code, the possibilities for DNA testing are seemingly endless.

Conclusion

DNA testing is a remarkable achievement of modern science, providing a window into our past, our biology, and our connections with others. By extracting and analyzing the fundamental building blocks of life, scientists can uncover answers to questions that range from "Where did my ancestors come from?" to "Do I carry a genetic risk for a specific disease?"

The process itself, though highly technical, is a testament to human ingenuity—taking us from a simple cheek swab to a deep understanding of our own unique genetic makeup. **"Behind the Scenes: Secrets of How Things Are Made"** peels back the layers of mystery surrounding DNA testing, revealing a process that is both complex and awe-inspiring. Whether used in medical diagnostics, crime-solving, or tracing your heritage, DNA testing continues to transform our understanding of the human experience.

Behind the Scenes

How Are Drones Designed and Operated?

Drones, also known as Unmanned Aerial Vehicles (UAVs), have become a crucial part of many industries, from military applications and aerial photography to package delivery and agriculture. But how are these flying marvels designed, and what makes them operate so effectively? The answer lies in a combination of cutting-edge engineering, advanced electronics, and precision software.

Designing a Drone: The Basics

At the core of drone design is the balance between weight, power, and control. Engineers need to ensure that drones are light enough to fly but strong enough to carry the necessary equipment, such as cameras, sensors, or cargo. The design process typically follows these main steps:

1. **Purpose and Functionality:** Every drone starts with a specific purpose, which heavily influences its design. For example, military drones may need long flight durations and stealth capabilities, while drones for photography prioritize stability and camera integration. The design must reflect the drone's

Behind the Scenes

intended use—whether it's for surveillance, mapping, or recreational purposes.

2. **Aerodynamics and Frame Design:** Aerodynamics plays a significant role in drone performance. The frame, often made from lightweight materials like carbon fiber or aluminum, is carefully designed to reduce air resistance while maintaining strength. The number of propellers, typically four in a quadcopter, ensures stability and allows for precise control of flight movements.

3. **Power Systems:** The power system is critical in drone design. Most drones are powered by lithium-polymer (Li-Po) batteries, which offer a high energy density, allowing drones to stay aloft for extended periods without adding excessive weight. The battery size and efficiency determine the drone's flight time, making power management a crucial design consideration.

4. **Propulsion System:** Propellers or rotors are vital for keeping the drone airborne and controlling its direction. Brushless motors are commonly used due to their efficiency, longevity, and reduced noise. These motors power the propellers, providing lift and enabling the drone to hover, accelerate, decelerate, and make turns mid-air.

5. **Sensors and Navigation:** Drones require a range of sensors to maintain stable flight. These include

Behind the Scenes

gyroscopes and accelerometers to detect movement, GPS modules for location tracking, and barometers to monitor altitude. Advanced drones also come equipped with obstacle detection sensors, such as infrared, ultrasonic, or lidar, allowing them to avoid collisions autonomously.

6. **Flight Control System:** The brain of the drone is its flight control system, typically a combination of hardware and software that interprets data from sensors and translates them into instructions for the motors. This system ensures stable flight, even in challenging weather conditions, and allows for remote control or automated flight paths.

7. **Communication Systems:** Drones rely on wireless communication systems to interact with their operators. Radio frequency (RF) communication, often in the 2.4 GHz or 5.8 GHz range, is commonly used to transmit signals from the controller to the drone. More sophisticated drones also employ real-time video feeds, streaming live footage back to the operator through Wi-Fi or cellular networks.

Operating a Drone

Once a drone is designed and built, operating it requires a combination of manual control and, in some cases, automated systems.

Behind the Scenes

1. **Remote Control Operation:** Most drones are operated using a remote controller, which allows the pilot to control its direction, speed, and altitude. These controllers often resemble a game console controller, with joysticks for movement and buttons for other functions, like adjusting the camera or taking off and landing. Pilots rely on visual line-of-sight (VLOS) to navigate the drone, although more advanced drones feature first-person view (FPV) systems, displaying a live feed from the drone's camera to the operator's screen.
2. **Autonomous Flight:** Many modern drones have the ability to fly autonomously, following a pre-programmed flight path or responding to real-time data. Using GPS and navigation software, the operator can plan a route, which the drone will then execute without needing continuous input. This feature is especially useful for tasks like surveying large areas or delivering packages.
3. **Flight Stabilization and Navigation:** Drones come equipped with flight stabilization software, which automatically adjusts the rotors to maintain a steady altitude and direction. This technology is particularly valuable in windy conditions or when precise hovering is required for photography or inspection tasks. GPS navigation helps the drone

Behind the Scenes

maintain its position and follow specific waypoints during autonomous missions.

4. **Obstacle Avoidance:** Many drones incorporate sophisticated obstacle detection and avoidance systems. Using sensors like lidar or cameras, drones can detect objects in their flight path and adjust course to avoid collisions. This feature is essential in applications like search and rescue or surveying, where navigating through complex environments is required.

5. **Landing Systems:** Drones often feature automated landing systems, which use GPS and sensor data to return the drone to its takeoff point or a designated landing area. These systems are critical when operating in challenging environments or when the operator loses sight of the drone. The "return-to-home" feature ensures that the drone safely lands if it loses connection with the controller or if the battery is low.

6. **Flight Regulations:** Operating a drone is subject to local and international regulations. In many countries, drone pilots need to register their drones and adhere to specific rules, such as flying below a certain altitude or avoiding restricted airspace, like near airports or military zones. For commercial operators, obtaining a license or certification is often

required, and strict safety guidelines must be followed.

Applications of Drones

The versatility of drones has opened up a wide array of applications across different industries:

- **Aerial Photography and Videography:** Drones have revolutionized the film and photography industry by providing affordable and dynamic aerial shots that were once only possible with helicopters.
- **Agriculture:** Farmers use drones to monitor crop health, manage irrigation, and even apply pesticides. These agricultural drones help increase efficiency and reduce resource consumption.
- **Delivery Services:** Companies like Amazon and UPS are experimenting with drone deliveries to improve logistics and reduce delivery times, particularly in remote areas.
- **Military and Surveillance:** Military drones are used for reconnaissance, target identification, and in some cases, direct attacks. These drones operate with advanced automation and are often capable of long-range, long-duration missions.
- **Search and Rescue:** Drones can access remote or dangerous areas more quickly than humans, helping in search and rescue missions after natural disasters,

monitoring forest fires, or delivering supplies to stranded individuals.
- **Environmental Monitoring:** Drones are employed to monitor wildlife, track deforestation, and collect data on environmental conditions, offering insights into climate change and its impact on ecosystems.

The Future of Drones

As technology advances, drones are set to become even more autonomous, intelligent, and versatile. Researchers are developing swarming technology, which allows multiple drones to work together collaboratively, as well as integrating artificial intelligence into drone operations, making them capable of performing complex tasks with minimal human intervention.

Drones will likely become an even more integral part of everyday life, transforming industries such as logistics, healthcare, and agriculture while continuing to push the boundaries of what's possible in exploration, surveillance, and automation.

In conclusion, the design and operation of drones represent the convergence of several advanced technologies, each contributing to the impressive capabilities these devices possess. Whether for recreation, business, or critical operations, drones are reshaping the way we interact with the world, giving us new perspectives and tools to solve problems more efficiently than ever before.

Behind the Scenes

How Is Glass Manufactured?

Glass is a material we encounter daily, from the windows in our homes and cars to the screens of our smartphones and the glasses we drink from. But how is this versatile and seemingly magical substance created? The process of glass manufacturing has evolved over millennia, combining science and artistry to produce one of the most important materials in human history. The method may seem mysterious, but it all starts with simple, natural ingredients transformed through heat and craftsmanship into the transparent, solid structures we rely on every day.

The Raw Materials

The primary ingredient in most glass is **silica**, or silicon dioxide (SiO_2), which is found in sand. Other key components are added to adjust the properties of the glass:

- **Soda ash (sodium carbonate)**: Lowers the melting point of silica, making the process more energy-efficient.
- **Limestone (calcium carbonate)**: Helps stabilize the mixture, making the final product more durable and resistant to water.
- **Alumina (aluminum oxide)**: Increases the glass's strength.

- **Various additives**: Depending on the intended use, materials such as iron oxide (for tinting), boron (for heat resistance), or lead (for crystal glass) may be included.

Melting: Turning Sand into Molten Glass

The first major step in glass manufacturing is **melting** the raw materials. This occurs in large furnaces at extremely high temperatures, typically around **1700°C (3090°F)**. Inside these furnaces, the mixture of silica, soda ash, and limestone is heated until it becomes a molten liquid. At this stage, the materials lose their crystalline structure and become an amorphous (non-crystalline) liquid state, which is the essence of glass.

The molten glass is intensely hot and viscous, much like thick honey. Achieving the right consistency is crucial since the viscosity must be just right for shaping.

Shaping the Glass

Once the molten glass reaches the desired consistency, it can be shaped in various ways depending on the product being made:

- **Blowing**: For items like bottles or decorative glass, glassblowers shape the molten glass by blowing air into a molten blob using a blowpipe. This is one of

the oldest techniques, used for centuries to create intricate designs.
- **Float Glass Process**: This is the most common method for making flat glass (used in windows, mirrors, etc.). In this process, molten glass is poured onto a bed of molten tin. Since glass and tin don't mix, the glass spreads out and floats on the tin's surface, forming a perfectly smooth sheet. The glass is then slowly cooled and solidified, forming large panes of flat glass.
- **Pressing and Molding**: In this method, molten glass is poured into molds to create objects like glasses, bowls, and lightbulbs. Machines press the glass into the mold, shaping it before it cools and solidifies.

Annealing: Cooling the Glass

After shaping, the glass must be cooled carefully in a process known as **annealing**. If glass cools too quickly, internal stresses can cause it to crack or shatter. To prevent this, the glass is slowly passed through an annealing oven (called a lehr), where it cools gradually over a period of hours or even days. This controlled cooling process relieves any stresses and makes the glass stronger and more stable.

Cutting and Finishing

Behind the Scenes

Once the glass has cooled and solidified, it undergoes further processing depending on its use:

- **Cutting**: Large sheets of glass are cut to specific sizes. This is especially important in industries like construction, where glass must fit precisely into window frames or buildings.
- **Polishing and Coating**: Glass for specialized applications may be polished to remove imperfections, and some may receive additional coatings. For example, **tempered glass** (used in car windows or phone screens) is treated with heat or chemicals to increase its strength, making it more resistant to breaking.
- **Tinting and Etching**: In some cases, glass is tinted or etched to add designs or provide additional functionality, such as UV protection for windows or decorative effects for glassware.

Specialty Glass and Innovations

Glass manufacturing has continued to evolve, with modern innovations producing specialized types of glass for various applications:

- **Tempered glass**: This type of glass undergoes a special heat treatment, making it up to four times stronger than standard glass. When it breaks, it shatters into small, blunt pieces rather than sharp

shards, making it safer for things like car windows and shower doors.
- **Laminated glass**: This type of glass consists of multiple layers, often with plastic interlayers, making it resistant to shattering. It's commonly used in windshields and building facades.
- **Fiberglass**: Created by pulling molten glass into fine threads, fiberglass is used for insulation and in the automotive and aerospace industries due to its lightweight and durable properties.
- **Smart glass**: This modern innovation can change its transparency at the touch of a button or in response to changes in light or temperature, making it useful for energy-efficient windows and high-tech displays.

Recycling Glass

One of the significant advantages of glass is that it's highly recyclable. Glass can be melted down and reformed endlessly without losing its quality, making it one of the most sustainable materials in use today. **Recycled glass**, known as **cullet**, is often mixed with new raw materials in the manufacturing process, reducing the energy required for production and helping minimize the environmental impact.

Conclusion

Behind the Scenes

The manufacturing of glass is a remarkable blend of ancient techniques and modern technology. From its origins as simple sand, through its transformation into molten liquid, to its final solid form, glass is an example of how humans have harnessed natural materials and developed processes to create something both practical and beautiful. Whether it's the windows that let light into our homes, the lenses that help us see, or the screens that connect us to the digital world, glass is a vital component of our everyday lives, and understanding how it's made only deepens our appreciation for this incredible material.

Behind the Scenes

How Are Bridges Constructed?

Bridges are among the most awe-inspiring feats of engineering, spanning rivers, valleys, highways, and even oceans to connect communities, transport goods, and facilitate travel. While they appear as static structures, the process of constructing a bridge is a complex and dynamic endeavor, requiring intricate planning, advanced technology, and a deep understanding of physics, materials science, and environmental conditions. Let's take a closer look at how these marvels of engineering are brought to life, from the initial design phase to the final ribbon-cutting ceremony.

1. Planning and Design

Before a single shovel of dirt is moved, extensive planning is required. The type of bridge to be constructed—whether it's a suspension bridge, arch bridge, beam bridge, or cable-stayed bridge—depends on several factors:

- **Geography:** The landscape determines much of the design. Engineers must assess the length of the span, the depth of the water or valley, and the stability of the surrounding ground.

- **Purpose:** Is the bridge meant for cars, trains, pedestrians, or a combination? The type of traffic affects the load-bearing requirements and the overall design.
- **Environmental Impact:** Engineers conduct studies to ensure that the bridge won't have a negative effect on local ecosystems, such as river flow or wildlife habitats.
- **Budget and Materials:** Financial constraints play a key role in determining the size and scope of the bridge, as well as the materials used (steel, concrete, composite materials).

During this phase, computer simulations are often used to model different bridge designs. These models allow engineers to predict how the bridge will react to various forces such as wind, earthquakes, or the weight of vehicles. Once a design is finalized, it moves into the next stage.

2. Site Preparation

With the bridge design in place, the next step is to prepare the construction site. This phase involves clearing the land, relocating any existing utilities (such as water or gas lines), and stabilizing the ground. For bridges over water, additional steps are necessary, including:

- **Cofferdams:** These temporary structures are built to create a dry work area within a body of water. Water is pumped out of the enclosed area, allowing workers to access the riverbed or seabed for foundation work.
- **Piling:** Large steel or concrete piles (long columns) are driven deep into the ground to provide a stable foundation. These piles can extend dozens or even hundreds of feet into the earth, reaching bedrock to ensure the bridge's stability.

3. Foundation Construction

A bridge's foundation is critical to its strength and durability. Foundations must be deep and sturdy enough to bear the immense weight of the bridge and the vehicles or pedestrians that will use it. The two main components of a bridge's foundation are:

- **Abutments:** These are the structures at either end of the bridge that support the ends of the span and anchor it to solid ground.
- **Piers:** Piers are vertical supports placed along the length of the bridge, usually in the middle of the span. For long bridges, multiple piers are placed at strategic points to support the load of the roadway.

For bridges over water, caissons (watertight retaining structures) are sometimes used to create a dry workspace at the bottom of a river or seabed, allowing workers to pour concrete and build the foundation under water.

4. Superstructure Construction

The superstructure is the part of the bridge that you see above ground. It includes the deck (the roadway or pathway that vehicles or pedestrians travel on) and the supporting structure that holds it in place. The construction of the superstructure varies based on the type of bridge:

- **Beam Bridges:** These are the simplest type of bridge, made up of horizontal beams supported by piers or abutments at either end. The beams can be constructed from steel, pre-stressed concrete, or other materials.
- **Arch Bridges:** The arch shape transfers the weight of the bridge and its load into the abutments. During construction, temporary scaffolding (known as centering) supports the arch until it is completed.
- **Suspension Bridges:** Suspension bridges, like the famous Golden Gate Bridge, are supported by cables suspended from tall towers. The cables are anchored at both ends of the bridge and carry the weight of the deck. To build a suspension bridge, towers are erected first, followed by the installation of the main cables. The deck is then hung from the cables in sections.

- **Cable-Stayed Bridges:** These bridges feature a deck that is directly supported by cables attached to one or more towers. The cables run directly from the towers to the deck, providing support without the need for additional piers or abutments in the middle.

5. Deck Installation

Once the main structural components are in place, the deck is installed. Depending on the bridge type, the deck might be built using steel, concrete, or a combination of materials. The deck serves as the driving or walking surface, so it must be constructed with durability and safety in mind. For long bridges, the deck is often built in sections, which are then lifted into place by cranes or other heavy machinery.

6. Finishing Touches and Safety Features

With the structure of the bridge complete, it's time to focus on the finishing touches. These include:

- **Guardrails and Barriers:** To protect drivers and pedestrians, guardrails or barriers are installed along the sides of the bridge.
- **Lighting:** Bridges often feature lighting systems for safety and aesthetic appeal. In some cases, decorative lighting can turn the bridge into a nighttime landmark.

Behind the Scenes

- **Expansion Joints:** Bridges are exposed to temperature changes that cause the materials to expand and contract. Expansion joints are placed between sections of the bridge to allow for these movements without damaging the structure.

Additionally, testing and inspections are performed to ensure the bridge meets all safety standards and regulations.

7. Opening and Maintenance

Once construction is complete, the bridge is opened to the public. But the work doesn't stop there—regular maintenance is essential to ensure the bridge remains safe and functional for years to come. Maintenance activities include:

- **Inspecting and repairing structural components:** Bridges are exposed to wear and tear from weather, traffic, and environmental factors. Inspections are conducted to identify any signs of damage, corrosion, or wear.
- **Repainting and resurfacing:** Steel bridges are often painted to protect against rust, while concrete bridges may need resurfacing to repair cracks or potholes.

Conclusion

Behind the Scenes

Bridge construction is a testament to human ingenuity, blending art, science, and engineering to overcome the challenges posed by nature and geography. Each bridge, whether a simple beam crossing a creek or a towering suspension bridge over a vast expanse, is a marvel of design and planning. The next time you cross a bridge, take a moment to appreciate the incredible effort, skill, and knowledge that went into making it possible—truly, an engineering masterpiece that connects us all.

Behind the Scenes

How is Data Encryption Done?

In our increasingly digital world, data encryption has become a vital process for protecting sensitive information. It serves as a robust defense mechanism against unauthorized access and cyber threats, safeguarding everything from personal messages to corporate data and government secrets. This chapter delves into the intricate world of data encryption, exploring how it works, the techniques used, and its significance in today's cybersecurity landscape.

Understanding Data Encryption

At its core, **data encryption** is the process of converting plaintext (readable data) into ciphertext (encoded data) to prevent unauthorized access. Only those who possess a specific key or password can decipher the ciphertext back into plaintext. This transformation ensures that even if data is intercepted, it remains unreadable to anyone without the correct decryption key.

The Importance of Encryption

Data encryption plays a crucial role in various domains, including:

- **Personal Security:** Individuals use encryption to protect sensitive information such as credit card numbers, social security numbers, and personal messages from hackers and identity thieves.
- **Corporate Protection:** Businesses encrypt data to safeguard proprietary information, customer records, and intellectual property, thus maintaining confidentiality and compliance with data protection regulations.
- **Communication Privacy:** Encryption is essential for secure communication, ensuring that emails, instant messages, and voice calls are protected from eavesdropping.
- **Data Integrity:** Encryption helps verify that data has not been altered or tampered with during transmission or storage.

How Encryption Works

Encryption relies on algorithms, which are mathematical formulas that dictate how data is transformed. There are two primary types of encryption algorithms: **symmetric encryption** and **asymmetric encryption**.

1. Symmetric Encryption

Symmetric encryption uses the same key for both encryption and decryption. This means that both the sender

Behind the Scenes

and the recipient must have access to the secret key, making it essential to protect this key from unauthorized access.

How It Works:

- **Key Generation:** A secret key is generated, typically through a random number generator.
- **Encryption Process:** The plaintext is processed through an encryption algorithm (e.g., AES, DES) using the secret key to produce ciphertext.
- **Decryption Process:** The recipient uses the same secret key and the decryption algorithm to convert the ciphertext back into plaintext.

Common Symmetric Encryption Algorithms:

- **AES (Advanced Encryption Standard):** Widely used for its security and speed, AES supports key sizes of 128, 192, and 256 bits.
- **DES (Data Encryption Standard):** An older algorithm that uses a 56-bit key, which is now considered less secure due to advancements in computing power.
- **RC4:** A stream cipher known for its speed but is less commonly used today due to vulnerabilities.

2. Asymmetric Encryption

Asymmetric encryption, also known as public-key cryptography, employs a pair of keys: a public key and a private key. The public key can be shared with anyone, while the private key is kept secret.

How It Works:

- **Key Pair Generation:** A pair of keys (public and private) is generated using an asymmetric algorithm (e.g., RSA, ECC).
- **Encryption Process:** The sender encrypts the plaintext using the recipient's public key, creating ciphertext that can only be decrypted with the corresponding private key.
- **Decryption Process:** The recipient uses their private key and the decryption algorithm to convert the ciphertext back into plaintext.

Common Asymmetric Encryption Algorithms:

- **RSA (Rivest-Shamir-Adleman):** One of the first widely used asymmetric algorithms, RSA relies on the difficulty of factoring large prime numbers.
- **ECC (Elliptic Curve Cryptography):** Offers similar security to RSA but with smaller key sizes, making it more efficient.

Additional Techniques in Data Encryption

Behind the Scenes

In addition to the basic encryption types, several techniques enhance data security:

1. Hashing

While not encryption in the traditional sense, hashing is a process that converts data into a fixed-size string of characters, which is unique to the original data. Hashing is commonly used to store passwords securely. Unlike encryption, hashing is one-way and cannot be reversed.

2. Salting

Salting adds random data (the salt) to the input of a hash function to ensure that identical inputs produce different hashes. This technique protects against dictionary and rainbow table attacks.

3. Key Management

Effective key management is crucial for encryption. Organizations must establish policies for generating, distributing, storing, and revoking encryption keys to prevent unauthorized access.

4. End-to-End Encryption (E2EE)

End-to-end encryption ensures that data is encrypted on the sender's device and only decrypted on the recipient's device, preventing intermediaries from accessing the data during transmission. This is commonly used in messaging applications like WhatsApp and Signal.

The Future of Data Encryption

As technology evolves, so do the threats to data security. Cybercriminals are continually developing new techniques to bypass encryption, making it essential for organizations and individuals to stay ahead of the curve. Innovations such as **quantum encryption** promise to enhance data protection significantly, using the principles of quantum mechanics to secure data transmission.

Conclusion

Data encryption is a vital component of modern cybersecurity, providing a robust defense against unauthorized access and protecting sensitive information across various sectors. Understanding how encryption works and the methods employed to secure data empowers individuals and organizations to make informed decisions about their digital security practices. In an era where data breaches and cyber threats are increasingly common, encryption remains one of the most effective tools in safeguarding our digital lives. By continuing to innovate and adapt, we can ensure that our data remains protected, secure, and accessible only to those who are authorized to view it.

Behind the Scenes

How is Virtual Reality Created?

Virtual reality (VR) has become a buzzword in the realms of gaming, entertainment, education, and various industries. It offers immersive experiences that transport users to different worlds, allowing them to interact with 3D environments in ways that traditional media cannot. But how is this groundbreaking technology created? Let's explore the process behind the creation of virtual reality, from the hardware components to the software development and design considerations.

1. Understanding the Components of Virtual Reality

Virtual reality is made up of several critical components that work together to create an immersive experience:

a. Hardware

1. **Head-Mounted Displays (HMDs):**
    - These are the primary devices used in VR, featuring screens placed in front of the user's eyes. They can be standalone devices or tethered to computers or gaming consoles.
    - Examples include the Oculus Quest, HTC Vive, and PlayStation VR, each with its own

specifications for resolution, refresh rate, and field of view.

2. **Input Devices:**
   - Controllers, gloves, and motion sensors track user movements and interactions within the VR environment.
   - Devices like the Oculus Touch controllers or Vive wand controllers provide haptic feedback and motion tracking to enhance realism.

3. **Sensors and Tracking Systems:**
   - Sensors are essential for tracking the user's head and body movements to adjust the VR experience in real-time.
   - Technologies such as infrared cameras, accelerometers, and gyroscopes are often utilized for precise tracking.

b. Software

1. Game Engines:
   - Popular game engines like Unity and Unreal Engine are commonly used for creating VR content. These engines provide the necessary tools to develop interactive 3D environments.
   - They offer built-in support for VR development, including optimized rendering and physics systems.

Behind the Scenes

2. **Development Platforms:**
    - VR development requires specific SDKs (Software Development Kits) and APIs (Application Programming Interfaces) to communicate between the hardware and software.
    - Examples include Oculus SDK, SteamVR, and OpenVR, which enable developers to build applications that leverage the unique features of VR hardware.

2. Designing the Virtual Environment

Creating a compelling virtual reality experience goes beyond simply putting users in a 3D world. The design of the virtual environment is crucial for immersion:

a. 3D Modeling and Animation:

- Artists and designers use software such as Blender, Maya, or 3ds Max to create the 3D models of characters, objects, and environments.
- Animations are essential for bringing life to these models, whether it's the movement of a character or the subtle sway of trees in the wind.

b. Sound Design:

- Audio is a vital aspect of VR that enhances immersion. Spatial audio technology allows sound

to be perceived from different directions, mimicking real-life experiences.
- Sound effects, ambient noises, and voiceovers are carefully integrated to provide a rich audio landscape.

c. User Interface (UI) and User Experience (UX):

- Designing intuitive interfaces for VR applications can be challenging. Developers must consider how users interact with the environment, ensuring controls and menus are easily accessible.
- The UX design must focus on creating a seamless and comfortable experience, avoiding motion sickness and disorientation.

3. Developing the VR Experience

The development phase involves several steps to bring the virtual reality experience to life:

a. Prototyping:

- Developers often create prototypes to test concepts and mechanics before full-scale development. This allows for early user feedback and iteration.

b. Programming:

Behind the Scenes

- Using scripting languages (such as C# for Unity or C++ for Unreal Engine), developers write the code that governs how the virtual environment behaves.
- This includes implementing physics, interactivity, and animations, ensuring that all elements work harmoniously.

c. Testing:

- Rigorous testing is essential to ensure that the VR experience is enjoyable and free from bugs. This involves checking for performance issues, tracking accuracy, and user comfort.
- User testing with diverse audiences can reveal insights into usability and immersion, allowing developers to make necessary adjustments.

4. Optimization and Finalization

Once the VR experience is built, optimization becomes crucial to ensure smooth performance:

a. Performance Optimization:

- VR applications require high frame rates to provide a smooth experience and prevent motion sickness. Developers may optimize graphics, reduce polygon counts, or implement level of detail (LOD) techniques to achieve this.

Behind the Scenes

b. Final Touches:

- The final stage involves polishing the application, adding details, and ensuring that all assets work together seamlessly.
- Quality assurance (QA) checks are performed to verify that everything functions as intended.

5. Deployment and Distribution

After the virtual reality experience is complete, it's time to share it with users:

a. Publishing:

- VR applications can be published on various platforms such as the Oculus Store, Steam, or PlayStation Store, depending on the target audience and hardware compatibility.

b. User Support:

- Developers often provide updates and support for their applications, addressing any issues and adding new content based on user feedback.

Conclusion

The creation of virtual reality is a complex, multi-faceted process that combines cutting-edge technology with artistry and engineering. From hardware components to

Behind the Scenes

software design, every aspect plays a crucial role in crafting immersive experiences that transport users to new worlds. As technology continues to evolve, the possibilities for virtual reality are boundless, promising even more innovative and captivating experiences in the future. Whether in gaming, education, healthcare, or beyond, virtual reality is revolutionizing the way we interact with information and environments, opening doors to endless exploration and discovery.

Behind the Scenes

How is Weather Modified?

Weather modification, often referred to as **geoengineering** or **weather control**, involves various techniques aimed at altering atmospheric conditions to achieve desired outcomes. While the idea of controlling the weather may sound like something out of science fiction, there are real-world methods employed for this purpose. The most common form of weather modification is cloud seeding, but there are also other techniques being explored. This chapter will delve into the science, methods, ethical considerations, and potential applications of weather modification.

Understanding Weather Modification

What Is Weather Modification?

Weather modification encompasses a range of techniques that aim to alter weather patterns. This can include increasing precipitation, reducing hail, or even influencing temperatures. The primary goal of weather modification is to mitigate the adverse effects of weather, enhance agricultural productivity, and manage water resources.

The Science Behind Weather Modification

Behind the Scenes

The fundamental concept behind weather modification lies in understanding the atmospheric processes that lead to precipitation. Clouds consist of tiny water droplets or ice crystals that form when moisture in the air cools and condenses. When these droplets coalesce into larger droplets, they fall as rain or snow. Weather modification seeks to enhance this natural process through various methods.

Common Methods of Weather Modification

1. Cloud Seeding

Cloud seeding is the most widely recognized method of weather modification. It involves dispersing substances into the atmosphere to encourage cloud condensation and precipitation. Common agents used in cloud seeding include:

- **Silver Iodide:** This compound has a structure similar to ice and is effective in encouraging ice crystal formation in cold clouds.
- **Sodium Chloride (Table Salt):** Salt particles can help attract moisture in warmer clouds.
- **Liquid Propane:** When aerosolized, liquid propane expands and cools, forming ice crystals in clouds.

Process of Cloud Seeding:

1. **Identification of Suitable Clouds:** Meteorologists assess cloud conditions to determine whether they are conducive to seeding. Ideal candidates are clouds with sufficient moisture but insufficient precipitation.
2. **Dispersal of Seeding Agents:** Aircraft equipped with flares or ground-based generators release the seeding agents into the atmosphere. The agents promote the formation of larger droplets, leading to increased precipitation.
3. **Monitoring:** Weather stations and satellites track changes in precipitation levels post-seeding to evaluate effectiveness.

2. Hail Suppression

Hail can cause significant damage to crops, property, and vehicles. Hail suppression techniques aim to reduce the size and frequency of hailstones. Cloud seeding can also be employed for this purpose, using similar techniques to encourage smaller ice crystals to form instead of larger hailstones.

3. Fog Dispersal

Fog can create dangerous driving conditions and disrupt airport operations. Fog dispersal techniques, such as the use of heated air or chemical agents, aim to clear fog by raising the temperature or altering the moisture content.

Behind the Scenes

For example, using a combination of hot air and water vapor can help dissipate fog.

4. Hurricane Modification

While still largely experimental, researchers have explored methods to influence hurricanes, including:

- **Cloud Seeding:** Attempting to weaken hurricanes by dispersing substances into the storm's clouds to promote more widespread rainfall.
- **Ocean Cooling:** Some studies suggest that cooling ocean surface temperatures could potentially reduce hurricane intensity. However, these methods are complex and fraught with ethical concerns.

Applications of Weather Modification

Agricultural Benefits

Weather modification has the potential to increase agricultural productivity by enhancing rainfall during dry periods. Farmers in arid regions have employed cloud seeding to ensure adequate water supply for crops, ultimately leading to better harvests.

Water Resource Management

In areas facing water shortages, weather modification can help replenish reservoirs and groundwater supplies. By increasing precipitation, cloud seeding can alleviate drought conditions and support sustainable water management.

Disaster Mitigation

Weather modification techniques can also play a role in disaster mitigation. For example, reducing hail size can protect crops and property, while fog dispersal can improve transportation safety.

Urban and Environmental Planning

Cities prone to drought or flooding can benefit from weather modification strategies as part of broader urban planning efforts. By managing precipitation patterns, city planners can better prepare for extreme weather events.

Ethical Considerations and Challenges

While the potential benefits of weather modification are significant, there are also ethical considerations and challenges to address:

1. Environmental Impact

The long-term environmental effects of weather modification techniques are still not fully understood. Altering natural weather patterns may have unintended consequences on ecosystems and local climates.

2. Equity and Access

Access to weather modification technologies may create inequalities, as wealthier regions may be able to afford such interventions while poorer areas may not. This raises questions about fair distribution of resources and benefits.

3. International Regulations

Weather modification can have cross-border implications. For example, increased rainfall in one area may lead to drought in another. Establishing international regulations and agreements is essential to prevent conflict over modified weather patterns.

4. Public Perception

Public perception of weather modification can vary widely. Some view it as a necessary tool for managing climate challenges, while others see it as an unethical manipulation of natural systems. Educating the public about the science and potential impacts is crucial.

Conclusion

Behind the Scenes

Weather modification represents a fascinating intersection of science, technology, and environmental stewardship. While cloud seeding and other techniques offer promising solutions to enhance precipitation, reduce natural disasters, and manage water resources, it is essential to approach these methods with caution. Understanding the science behind weather modification and addressing ethical considerations will be key to harnessing its potential for the benefit of society.

As we continue to face the challenges of climate change and extreme weather events, the exploration of weather modification techniques may provide valuable tools in our quest for sustainable solutions. The journey of understanding how weather is modified invites us to appreciate the delicate balance of nature and the innovative spirit of humanity in our efforts to adapt and thrive.

Behind the Scenes

How Are Autonomous Vehicles Programmed?

Autonomous vehicles, commonly known as self-driving cars, represent one of the most significant advancements in transportation technology. These vehicles rely on a complex blend of hardware and software to navigate, make decisions, and operate safely without human intervention. Understanding how autonomous vehicles are programmed requires delving into a multitude of disciplines, including computer science, robotics, machine learning, and artificial intelligence (AI). This chapter will explore the intricate processes involved in programming autonomous vehicles, highlighting the key technologies, methodologies, and challenges that developers face.

1. Core Components of Autonomous Vehicles

Before diving into programming specifics, it's essential to recognize the core components that enable autonomous vehicles to function:

- **Sensors:** Autonomous vehicles are equipped with a variety of sensors, including LiDAR (Light Detection and Ranging), cameras, radar, and ultrasonic sensors. These devices help the vehicle perceive its environment by detecting obstacles,

road signs, lane markings, pedestrians, and other vehicles.
- **Computing Hardware:** Powerful onboard computers process data from the sensors in real time. These systems use advanced algorithms to interpret the information, enabling the vehicle to make informed decisions based on its surroundings.
- **Control Systems:** Control systems are responsible for managing the vehicle's movements, including steering, acceleration, and braking. These systems ensure the vehicle responds appropriately to its environment, making safe maneuvers based on programmed logic.

2. Data Collection and Preparation

The first step in programming an autonomous vehicle is gathering vast amounts of data. This data is crucial for training machine learning models that enable the vehicle to recognize and respond to various driving scenarios.

- **Driving Scenarios:** Developers collect data from real-world driving experiences, simulating different environments, weather conditions, and traffic situations. This data helps to create a comprehensive database of driving scenarios that the vehicle may encounter.

- **Annotation:** The collected data must be annotated to provide context for the machine learning models. For instance, labeling images from cameras to identify pedestrians, vehicles, traffic signs, and road markings helps the algorithm learn to recognize these elements in real-time.

3. Machine Learning and AI Algorithms

At the heart of autonomous vehicle programming is machine learning, particularly deep learning, which allows the vehicle to learn from data rather than being explicitly programmed for every scenario.

- **Neural Networks:** Autonomous vehicles typically employ convolutional neural networks (CNNs) to process visual data from cameras. These networks are designed to identify and classify objects in images, enabling the vehicle to understand its environment.
- **Reinforcement Learning:** This approach allows the vehicle to learn optimal driving strategies through trial and error. The vehicle receives feedback based on its actions (reward for safe driving, penalties for collisions), which helps refine its decision-making processes over time.

4. Path Planning and Decision-Making

Once the vehicle perceives its environment and identifies objects, it must plan a safe path and make decisions on how to navigate.

- **Path Planning Algorithms:** These algorithms calculate the best route for the vehicle to follow, considering factors like traffic rules, road conditions, and the behavior of surrounding vehicles. Techniques such as A* search, Dijkstra's algorithm, and Rapidly-exploring Random Trees (RRT) are commonly used.
- **Behavior Prediction:** To navigate safely, the vehicle must predict the behavior of other road users. This involves analyzing the movement patterns of pedestrians, cyclists, and other vehicles to anticipate their actions and avoid potential collisions.

5. Control Systems Integration

After determining a path and predicting behaviors, the vehicle's control systems come into play to execute the planned maneuvers.

- **Feedback Loops:** Autonomous vehicles use feedback loops to continuously adjust their actions based on real-time data. For example, if the vehicle detects a sudden obstacle, the control system will adjust the speed or direction accordingly.

Behind the Scenes

- **Testing and Calibration:** Extensive testing is vital to ensure that the control systems respond accurately and safely under various conditions. Developers simulate different scenarios and calibrate the vehicle's response to ensure optimal performance.

6. Simulation and Testing

Before deploying autonomous vehicles on public roads, extensive simulation and testing are essential to validate the programming.

- **Simulated Environments:** Developers use simulation software to create virtual environments where they can test the vehicle's algorithms without risk. These simulations can replicate a wide range of driving conditions, allowing for thorough testing of the vehicle's decision-making processes.
- **Real-World Testing:** Once simulations are complete, autonomous vehicles undergo real-world testing to evaluate their performance in genuine driving scenarios. This step is critical for identifying any issues that may not have been apparent in simulations.

7. Safety and Regulatory Compliance

Safety is paramount in the development of autonomous vehicles. Developers must adhere to stringent safety standards and regulations to ensure that their vehicles operate reliably in public spaces.

- **Safety Protocols:** Manufacturers implement rigorous safety protocols, including redundant systems that can take over if one component fails. For instance, if a sensor malfunctions, another sensor can provide the necessary data to maintain safe operation.
- **Regulatory Approval:** Autonomous vehicles must comply with local and national regulations. This includes meeting specific safety criteria and passing evaluations conducted by regulatory bodies before they can be introduced into public traffic.

8. The Future of Autonomous Vehicle Programming

As technology continues to advance, the programming of autonomous vehicles will evolve. Future developments may include enhanced AI capabilities, improved sensor technologies, and more robust data-sharing systems among vehicles.

- **Vehicle-to-Everything (V2X) Communication:** This emerging technology allows autonomous vehicles to communicate with each other and with infrastructure, such as traffic lights and road signs, to improve navigation and safety.

- **Continuous Learning:** Future autonomous vehicles may be designed to learn continuously from their experiences, adapting to new scenarios in real-time and improving their algorithms over time.

Conclusion

The programming of autonomous vehicles is a complex and multidisciplinary endeavor that combines data science, machine learning, engineering, and regulatory compliance. As we have explored, it involves an intricate dance of perception, decision-making, and action—ensuring that these vehicles can navigate the world safely and efficiently. As technology progresses, the dream of fully autonomous vehicles operating seamlessly alongside human-driven cars is steadily becoming a reality, revolutionizing transportation for future generations.

Behind the Scenes

How is Coffee Processed from Bean to Cup?

Coffee is one of the most beloved beverages worldwide, enjoyed for its rich flavor and stimulating properties. But have you ever wondered what goes into bringing that perfect cup of coffee to your table? The journey from bean to cup is a complex process involving meticulous cultivation, harvesting, processing, roasting, and brewing. This guide will take you through each step, revealing the intricate details behind one of the most popular drinks in the world.

1. Cultivation

The coffee journey begins in the lush, tropical climates of coffee-growing regions, primarily found between the Tropics of Cancer and Capricorn. The two most widely cultivated species are **Coffea arabica** (Arabica) and **Coffea canephora** (Robusta).

- **Growing Conditions:** Coffee plants thrive at high altitudes, typically between 2,000 and 6,000 feet above sea level. They require well-drained soil, consistent rainfall, and protection from direct sunlight.

- **Planting:** Coffee seeds are planted in shaded nurseries, where they germinate and grow into young coffee plants. Once they reach about six months of age, they are transferred to larger fields.

2. Harvesting

Harvesting coffee cherries is a labor-intensive process that occurs once the cherries have ripened, typically 7 to 9 months after flowering.

- **Ripeness:** Coffee cherries change color from green to bright red when ripe, signaling that they are ready for harvesting.
- **Methods:** There are two main methods of harvesting:
    - **Selective Picking:** Skilled workers hand-pick only the ripe cherries, ensuring higher quality. This method is labor-intensive but yields the best flavor.
    - **Strip Picking:** Workers strip all cherries from the branches, regardless of ripeness. This method is faster but often results in lower-quality coffee due to the inclusion of unripe and overripe cherries.

3. Processing

Once harvested, the cherries must be processed quickly to prevent spoilage. There are two primary methods for processing coffee cherries: the **wet method** and the **dry method**.

- **Wet Processing:**
    - The cherries are pulped to remove the outer skin and then fermented in water to break down the mucilage surrounding the beans.
    - After fermentation, the beans are washed, dried in the sun or mechanical dryers, and hulled to remove any remaining layers of parchment.
- **Dry Processing:**
    - The cherries are spread out in the sun to dry naturally. Once dried, the outer layers are mechanically removed.
    - This method is less water-intensive and can enhance fruity flavors but is more prone to defects if not monitored carefully.

4. Milling

After processing, the green coffee beans must be milled to prepare them for export and roasting.

- **Hulling:** The parchment layer surrounding the beans is removed, resulting in green coffee beans.

- **Grading and Sorting:** Beans are graded based on size, weight, and quality. They are also sorted to remove any defective beans, ensuring uniformity in the final product.

5. Roasting

Roasting transforms green coffee beans into the aromatic, flavorful beans we recognize.

- **Roasting Process:**
    - Beans are heated in a roasting machine at temperatures ranging from 350°F to 500°F. The duration of roasting affects flavor, aroma, and color.
    - During roasting, the beans undergo a series of chemical reactions known as the **Maillard reaction** and **caramelization**, which develop the distinct flavors and aromas of coffee.
- **Roast Levels:** Coffee can be roasted to different levels: light, medium, or dark, each offering a unique taste profile. Light roasts tend to be more acidic and fruity, while dark roasts are richer and bolder.

6. Grinding

Once roasted, coffee beans are cooled and then ground to prepare them for brewing.

- **Grind Size:** The grind size can vary depending on the brewing method—coarse for French press, medium for drip coffee makers, and fine for espresso machines. The grind size affects the extraction process and ultimately the flavor of the coffee.

7. Brewing

Brewing is the final step in the coffee-making process, where hot water extracts flavors from the ground coffee.

- **Brewing Methods:** There are numerous methods for brewing coffee, each influencing flavor and strength:
    - **Drip Brewing:** Hot water passes through coffee grounds in a filter, resulting in a clean cup.
    - **French Press:** Coarse coffee grounds steep in hot water before being pressed, producing a fuller-bodied coffee.
    - **Espresso:** Hot water is forced through finely-ground coffee under pressure, creating a concentrated shot.
    - **Cold Brew:** Coarse coffee is steeped in cold water for an extended period, yielding a smooth, less acidic flavor.

8. Enjoying the Final Product

Behind the Scenes

Finally, your cup of coffee is ready to be enjoyed! Whether served black or adorned with milk and sugar, the flavors and aromas are a culmination of a meticulous process that began with a simple seed.

Conclusion

The journey from bean to cup is a remarkable testament to the art and science of coffee production. Each step, from cultivation to brewing, is a carefully orchestrated process that reflects the dedication of farmers, processors, and roasters worldwide. The next time you savor your cup of coffee, take a moment to appreciate the rich history and intricate steps that have brought this beloved beverage to your table.

Behind the Scenes

How Are Volcanoes Studied?

Volcanoes are among the most powerful and awe-inspiring natural phenomena on Earth. Their explosive potential and ability to reshape landscapes pose significant risks to nearby populations, ecosystems, and infrastructure. Understanding how volcanoes work and predicting their behavior is crucial for mitigating these risks. Scientists employ a variety of methods and tools to study volcanoes, integrating field observations, laboratory analyses, and advanced technologies to gain insights into these geological wonders.

1. Field Studies and Observation

Fieldwork is a fundamental component of volcanology. Researchers often conduct detailed field studies to observe volcanoes firsthand. This includes:

- **Mapping and Sampling:** Geologists create detailed maps of volcanic landforms and deposits. They collect rock samples from various parts of the volcano to analyze their composition, age, and eruption history. Understanding the physical characteristics of volcanic rocks helps scientists deduce the types of eruptions that have occurred.

Behind the Scenes

- **Monitoring Active Volcanoes:** Scientists use various instruments to monitor volcanic activity. They observe signs of unrest, such as increased seismic activity, ground deformation, gas emissions, and changes in temperature. For example, seismographs measure the frequency and intensity of earthquakes that may indicate magma movement beneath the surface.

2. Remote Sensing Technologies

Advancements in technology have revolutionized the way volcanoes are studied. Remote sensing techniques allow scientists to gather data from a safe distance and obtain information that may not be visible from the ground:

- **Satellite Imagery:** Satellites equipped with high-resolution cameras and sensors capture images of volcanoes over time. These images help researchers track changes in landforms, thermal activity, and ash plumes during eruptions.
- **LiDAR (Light Detection and Ranging):** LiDAR uses laser pulses to create detailed three-dimensional maps of volcanic landscapes. This technology is invaluable for monitoring surface changes caused by eruptions, such as lava flows and ash deposits.

- **InSAR (Interferometric Synthetic Aperture Radar):** This technique uses radar signals from satellites to detect subtle ground movements associated with volcanic activity. InSAR can identify deformation patterns that indicate magma accumulation or withdrawal.

3. Geochemistry and Gas Emissions

Analyzing the chemical composition of volcanic rocks and gases provides critical insights into volcanic processes:

- **Rock Analysis:** Scientists study the mineralogy and geochemistry of volcanic rocks to understand their formation and evolution. Techniques like X-ray fluorescence (XRF) and mass spectrometry allow researchers to determine the elemental composition of samples.
- **Gas Emissions:** Volcanoes release gases such as sulfur dioxide (SO_2), carbon dioxide (CO_2), and water vapor during eruptions and non-eruptive phases. Monitoring gas emissions helps scientists assess volcanic activity and potential hazards. For example, an increase in SO_2 emissions may indicate rising magma.

4. Seismology and Ground Deformation

Behind the Scenes

Seismic monitoring is essential for understanding volcanic processes. By analyzing seismic waves generated by earthquakes, scientists can infer the movement of magma and tectonic activity:

- **Seismographs:** Networks of seismometers are deployed around volcanoes to record seismic activity. These instruments detect vibrations in the ground, allowing researchers to identify patterns that precede eruptions.
- **Ground Deformation Monitoring:** Techniques such as GPS and tiltmeters measure ground deformation around volcanoes. Swelling or subsidence of the ground can indicate magma movement, providing crucial information about potential eruptions.

5. Volcanic Modeling and Simulation

Computer modeling plays a vital role in predicting volcanic behavior and understanding complex processes:

- **Eruption Forecasting:** Scientists use numerical models to simulate volcanic eruptions and assess potential impacts. These models incorporate data on magma composition, pressure, temperature, and gas content to predict eruption styles and hazards.

- **Hazard Assessments:** Volcanologists use simulations to create hazard maps, identifying areas at risk from lava flows, ashfall, and pyroclastic flows. These maps are crucial for emergency planning and community preparedness.

6. Collaboration and Public Outreach

Studying volcanoes is a multidisciplinary effort that involves collaboration among scientists, government agencies, and local communities:

- **Interdisciplinary Teams:** Volcanologists often work with geologists, geochemists, seismologists, and other experts to gain a comprehensive understanding of volcanic systems. This collaboration enhances the accuracy of predictions and risk assessments.
- **Community Engagement:** Educating local communities about volcanic hazards is essential for preparedness and safety. Scientists often conduct outreach programs to raise awareness and share information about monitoring efforts and potential risks.

Conclusion

Behind the Scenes

The study of volcanoes is a dynamic and multifaceted field that combines traditional geological methods with cutting-edge technology. Through field observations, remote sensing, geochemical analysis, seismic monitoring, and modeling, scientists strive to unlock the mysteries of these powerful natural phenomena. Understanding how volcanoes behave not only enhances our knowledge of Earth's geology but also plays a crucial role in protecting lives and property from volcanic hazards. As research continues to evolve, our ability to predict volcanic eruptions and mitigate their impacts will improve, ensuring a safer future for communities living in the shadow of these majestic giants.

Behind the Scenes

How is Currency Printed and Secured?

Currency is more than just a medium of exchange; it is a symbol of trust, a representation of value, and a vital component of a nation's economy. The process of printing and securing currency involves a complex interplay of technology, design, and security measures to ensure its integrity and prevent counterfeiting. This chapter will explore the intricate steps involved in currency production, from the initial design concepts to the final security features that keep our money safe.

1. The Design Process

The journey of currency begins with a meticulous design process. Central banks commission artists and designers to create banknotes that reflect the nation's history, culture, and values. Key elements of the design process include:

- **Imagery and Symbolism:** Each currency note features significant historical figures, landmarks, and symbols that represent the nation's identity. For example, the U.S. dollar prominently displays founding fathers, while other countries might showcase local heroes or cultural icons.

- **Layout and Color:** Designers choose colors and layouts that enhance the visual appeal while ensuring that the notes are distinguishable from one another. Different denominations often have unique color schemes to facilitate recognition.
- **Security Features:** Security is paramount, and designers incorporate various elements that are difficult to replicate. These features include watermarks, security threads, and microprinting, which are embedded during the design phase.

2. Material Selection

Currency is not printed on ordinary paper; it requires specialized materials that provide durability and security. Common materials include:

- **Cotton and Linen Blends:** Many countries use a cotton-linen blend for their banknotes, offering a soft, durable texture that can withstand wear and tear.
- **Polymer:** An increasing number of nations are adopting polymer banknotes, which are more durable, resistant to dirt and moisture, and less likely to be counterfeited due to the difficulty of replicating the material.

3. Printing Techniques

Behind the Scenes

Once the design and materials are finalized, the actual printing process begins. Various advanced techniques are employed, including:

- **Offset Printing:** This method is commonly used for the background colors and patterns on banknotes. It allows for high-quality imagery and can produce complex designs.
- **Intaglio Printing:** This technique is essential for producing the raised elements on banknotes, such as portraits and text. Intaglio involves engraving designs onto a metal plate, which is then inked and pressed onto the currency paper. This creates the distinctive texture that can be felt when running fingers over the note.
- **Digital Printing:** Some central banks are beginning to incorporate digital printing technology, which offers flexibility and the ability to create customized designs quickly.

4. Adding Security Features

To protect against counterfeiting, currency notes are embedded with a variety of sophisticated security features, which include:

- **Watermarks:** These are recognizable images embedded into the banknote during the papermaking process. They are visible when held up to the light

Behind the Scenes

and can be an essential factor in authenticating currency.

- **Security Threads:** These are thin strips of plastic or metallic material that are embedded within the note. They may appear as a solid line or as a series of micro-printed text when viewed under specific angles.
- **Holograms:** Some modern currencies feature holographic images that change appearance when tilted. These complex graphics are difficult to reproduce and provide a significant security advantage.
- **Color-Shifting Ink:** Certain denominations use ink that changes color when viewed from different angles. This feature adds another layer of complexity for potential counterfeiters.

5. Quality Control

Once printed, each batch of banknotes undergoes rigorous quality control processes. These checks ensure that the notes meet the strict standards for color, clarity, and security features. Any defects or inconsistencies result in the destruction of those notes, preventing them from entering circulation.

6. Distribution and Circulation

Behind the Scenes

After quality control, the finished currency is packaged and distributed to banks and financial institutions. Central banks carefully monitor the circulation of currency to maintain its integrity and prevent excess cash from flooding the market, which can lead to inflation.

7. Counterfeit Prevention and Education

To combat counterfeiting, central banks and governments invest in public education initiatives. They provide citizens with information on how to identify authentic currency by promoting awareness of the various security features. This education empowers individuals to recognize counterfeits and helps to maintain confidence in the currency system.

8. Continuous Improvement and Innovation

As technology evolves, so too do the methods for printing and securing currency. Central banks are constantly researching and implementing new technologies to stay ahead of counterfeiters. Innovations such as advanced biometric security features and improved digital currency systems are on the horizon, promising to enhance the security and efficiency of money in the future.

Conclusion

The process of printing and securing currency is a remarkable blend of artistry, science, and technology.

Behind the Scenes

From the initial design concepts to the intricate security features that protect our money, every step is crucial to maintaining trust in the currency system. Understanding how currency is made not only highlights the complexity behind something we often take for granted, but it also reinforces the importance of vigilance in preserving the integrity of our monetary systems.

Behind the Scenes

How Are Smartphones Manufactured?

Smartphones have become an integral part of our daily lives, acting as personal assistants, communication devices, entertainment hubs, and much more. But have you ever wondered how these complex devices are manufactured? The process is a remarkable blend of advanced technology, precise engineering, and efficient logistics. In this chapter, we will explore the various stages of smartphone manufacturing, from design to production and assembly, providing a comprehensive understanding of how these incredible devices come to life.

1. Design and Development

The manufacturing process begins long before the first smartphone rolls off the assembly line. It starts with a concept and design phase, where product designers and engineers collaborate to create a blueprint for the new device. This stage involves:

- **Market Research:** Manufacturers analyze consumer trends, preferences, and technological advancements to identify features that will appeal to users.

Behind the Scenes

- **Prototyping:** Once the design concept is established, engineers create prototypes to test functionality, ergonomics, and user experience. These prototypes undergo rigorous testing to refine the design.
- **Component Selection:** Designers choose materials and components, such as processors, memory chips, screens, and cameras, considering factors like performance, cost, and compatibility.

2. Sourcing Materials

Once the design is finalized, the next step involves sourcing the materials and components needed for production. This stage requires a complex supply chain that spans the globe, as different components are manufactured in various countries. Key aspects include:

- **Raw Materials:** Essential materials like aluminum, glass, plastics, and rare earth metals are sourced. For instance, lithium is crucial for batteries, while cobalt and nickel are important for producing durable phone bodies.
- **Component Manufacturers:** Major components, such as processors (e.g., Qualcomm, Apple's A-series chips), displays (e.g., OLED screens), cameras (e.g., Sony sensors), and batteries (e.g.,

lithium-ion), are produced by specialized manufacturers.

3. Component Manufacturing

After sourcing materials, individual components are manufactured. Each component undergoes specific manufacturing processes, including:

- **Chip Production:** Semiconductor manufacturers fabricate processors and memory chips using intricate processes that involve photolithography, etching, and doping silicon wafers.
- **Display Manufacturing:** Screens are produced by layering thin-film transistors, liquid crystals, and backlighting technology. This process requires precise calibration to ensure high resolution and color accuracy.
- **Camera Assembly:** Camera modules are assembled using precision optics and sensors, often involving multiple lenses and image stabilization technologies.

4. Assembly

Once all the components are ready, the assembly process begins. This stage typically occurs in large factories, often located in regions like China, where skilled labor is available. Key steps include:

- **Production Lines:** Smartphones are assembled on fast-paced production lines, where workers and machines work in sync to put together the device. Each worker is usually responsible for a specific task, ensuring efficiency.
- **Quality Control:** Throughout the assembly process, quality control checks are conducted to identify defects. These checks may involve visual inspections, functional testing, and automated systems that monitor performance.
- **Soldering and Connectivity:** Components are soldered onto the smartphone's main circuit board. This step ensures that all parts, including the processor, memory, and sensors, are correctly connected and functional.

5. Software Installation

While hardware is being assembled, software engineers work on the operating system and applications. This involves:

- **Operating System Installation:** The smartphone's operating system (e.g., Android, iOS) is installed on the device, along with any pre-loaded applications and firmware.
- **Quality Assurance:** Software testing is conducted to ensure that all applications run smoothly and that

the device functions as intended. This includes stress testing, performance assessments, and compatibility checks.

6. Packaging and Shipping

Once assembly and testing are complete, the smartphones are prepared for packaging and distribution:

- **Packaging Design:** The final product is packaged with protective materials, user manuals, and accessories (like chargers and earphones) to ensure safe transportation and a positive unboxing experience.
- **Logistics and Distribution:** After packaging, smartphones are shipped to distributors, retailers, and consumers worldwide. This phase involves sophisticated logistics and inventory management systems to track shipments and manage supply chains.

7. After-Sales Support

The manufacturing process doesn't end with the sale. After-sales support is crucial for maintaining customer satisfaction:

Behind the Scenes

- **Customer Service:** Manufacturers provide customer support for issues like software updates, hardware repairs, and troubleshooting.
- **Warranty and Repairs:** Most smartphones come with warranties that cover defects and malfunctions, prompting manufacturers to establish repair centers and service networks.

Conclusion

The manufacturing of smartphones is a complex and multifaceted process that combines cutting-edge technology with meticulous craftsmanship. From initial design and sourcing materials to assembly, software installation, and distribution, every step is critical in bringing these remarkable devices to life. As technology continues to evolve, so too will the manufacturing processes, leading to even more advanced and efficient smartphones in the future. Understanding how smartphones are made not only enhances our appreciation for these devices but also highlights the incredible innovations that drive our modern world.

Behind the Scenes

How is Food Preserved?

Food preservation is an age-old practice that has evolved significantly over the centuries. The fundamental purpose of food preservation is to extend the shelf life of food, maintaining its quality, nutritional value, and safety for consumption. As societies developed, so did their methods for keeping food from spoiling. Today, we rely on a variety of techniques that leverage science, technology, and traditional knowledge to ensure our food remains safe and enjoyable to eat. This chapter explores the various methods of food preservation, their mechanisms, and the science behind them.

The Importance of Food Preservation

Before diving into the methods, it's essential to understand why food preservation is vital. Food spoilage can occur due to several factors, including:

- **Microbial Growth:** Bacteria, yeasts, and molds thrive in moist environments and can rapidly multiply, causing food to spoil.
- **Enzymatic Activity:** Natural enzymes within fruits and vegetables can lead to deterioration in color, texture, and flavor.
- **Oxidation:** Exposure to oxygen can cause food to become stale or rancid, especially in fats and oils.

Behind the Scenes

By employing preservation techniques, we can:

- Reduce food waste by extending the lifespan of perishable items.
- Enhance food safety, decreasing the risk of foodborne illnesses.
- Allow for seasonal foods to be enjoyed year-round.
- Conserve nutrients, keeping the food's beneficial properties intact.

Common Methods of Food Preservation

1. **Canning**
    - **Process:** Canning involves sealing food in airtight containers and heating them to destroy harmful microorganisms. The heat also creates a vacuum seal that prevents new bacteria from entering.
    - **Types:** Pressure canning (for low-acid foods like vegetables and meats) and water bath canning (for high-acid foods like fruits and pickles) are common techniques.
    - **Benefits:** Canned foods can last for years and retain most of their nutritional value.
2. **Freezing**
    - **Process:** Freezing slows down enzyme activity and microbial growth by reducing the

temperature of food to below freezing (32°F or 0°C).
- **Techniques:** Foods are typically blanched (briefly boiled) before freezing to inactivate enzymes, then stored in airtight containers or vacuum-sealed bags.
- **Benefits:** Freezing preserves the texture and flavor of food, and it allows for long-term storage without significant loss of nutrients.

3. **Dehydration**
   - **Process:** Dehydration removes moisture from food, inhibiting the growth of bacteria and mold. This can be done using air drying, sun drying, or through mechanical dehydrators.
   - **Applications:** Commonly used for fruits, vegetables, and meats (jerky).
   - **Benefits:** Dried foods are lightweight, have a long shelf life, and retain concentrated flavors and nutrients.

4. **Fermentation**
   - **Process:** Fermentation utilizes beneficial bacteria or yeasts to convert sugars into acids, gases, or alcohol. This method not only preserves food but also enhances its flavor and nutritional profile.
   - **Examples:** Common fermented foods include yogurt, sauerkraut, kimchi, and pickles.

- **Benefits:** Fermented foods are often rich in probiotics, which promote gut health.

5. **Pickling**
    - **Process:** Pickling involves immersing food in an acidic solution, typically vinegar, or in a brine solution (saltwater). The acid or salt inhibits the growth of spoilage organisms.
    - **Types:** Quick pickling (refrigerated) and long-term pickling (processed in jars for shelf stability) are popular methods.
    - **Benefits:** Pickled foods have a tangy flavor and can last for several months to years.

6. **Smoking**
    - **Process:** Smoking imparts flavor to food while also preserving it. The smoke contains compounds that inhibit bacterial growth, and the process also dries out the food.
    - **Applications:** Commonly used for fish, meats, and cheeses.
    - **Benefits:** Smoked foods have a unique flavor profile and can last longer than their fresh counterparts.

7. **Vacuum Sealing**
    - **Process:** Vacuum sealing removes air from packaging, creating a tight seal that reduces oxidation and slows spoilage.

- **Applications:** This method is often used in conjunction with other preservation techniques like freezing or sous vide cooking.
- **Benefits:** Vacuum-sealed foods retain freshness and flavor, preventing freezer burn and extending shelf life.

8. **Irradiation**
   - **Process:** Food irradiation uses ionizing radiation to kill bacteria, parasites, and insects, extending the shelf life of foods. This method also slows down ripening and sprouting.
   - **Applications:** Commonly applied to spices, dried fruits, and some meats.
   - **Benefits:** Irradiated foods maintain their nutritional value and are considered safe by many health organizations.

9. **Sugar Preservation**
   - **Process:** Sugar acts as a preservative by drawing moisture out of food through osmosis. This is commonly used in making jams, jellies, and candied fruits.
   - **Benefits:** Foods preserved with sugar can have a longer shelf life and maintain sweetness and flavor.

10. **Salting**

- **Process:** Salting preserves food by drawing out moisture and creating an environment that is inhospitable to bacteria. This method is commonly used for meats and fish.
- **Benefits:** Salted foods, such as cured meats, can last for months without refrigeration.

Conclusion

Food preservation techniques have transformed the way we consume and enjoy food. Each method has its unique processes and benefits, making it possible for us to savor seasonal flavors throughout the year while reducing waste and ensuring safety. As we continue to explore innovative ways to preserve food in a rapidly changing world, understanding these techniques becomes increasingly important for making informed choices about what we eat. Whether you are canning your garden harvest, freezing leftovers, or fermenting vegetables, the art of food preservation connects us to our past while paving the way for a sustainable future.

How is Oil Extracted and Refined?

Oil is one of the most vital resources on the planet, powering our vehicles, heating our homes, and serving as a key ingredient in countless products. The journey of oil from the depths of the earth to your everyday use involves complex processes of extraction and refining. This chapter will delve into how oil is extracted and refined, shedding light on the techniques, technologies, and environmental considerations involved in bringing this essential resource to market.

The Extraction Process

1. Exploration

The first step in oil extraction is exploration, which involves identifying potential oil reserves. Geologists use various methods to locate underground reservoirs, including:

- **Seismic Surveys:** This technique involves sending sound waves into the ground. By analyzing the waves that bounce back, scientists can map the geological structures beneath the earth's surface and identify potential oil deposits.

Behind the Scenes

- **Geological Studies:** Researchers study rock formations, surface geology, and the history of an area to determine where oil might be located.

Once a potential site is identified, exploratory drilling is conducted to confirm the presence of oil.

2. Drilling

After confirming the presence of oil, drilling operations commence. There are two primary types of drilling:

- **Conventional Drilling:** This involves vertical drilling into the earth to reach oil reservoirs. Once the well reaches the oil formation, the pressure often causes the oil to flow naturally to the surface. However, this is not always sufficient, and additional techniques may be required.
- **Directional Drilling:** This method allows for drilling at various angles, often to access oil deposits that are not directly below the drilling site. This is especially useful in urban areas or offshore drilling.

3. Production

Once the well is drilled, the production phase begins. The extracted oil is often accompanied by natural gas, water, and other substances. Techniques used to enhance oil recovery include:

- **Primary Recovery:** This relies on the natural pressure of the reservoir to push oil to the surface. Typically, this method recovers only 10-20% of the oil in a reservoir.
- **Secondary Recovery:** When natural pressure decreases, water or gas is injected into the reservoir to maintain pressure and force more oil to the surface. This method can recover an additional 20-30% of the oil.
- **Tertiary Recovery (Enhanced Oil Recovery):** This involves advanced techniques like thermal recovery (injecting steam) or chemical flooding (injecting chemicals to reduce oil viscosity) to extract even more oil.

The Refining Process

Once oil is extracted, it undergoes refining to transform it into usable products. The refining process can be broken down into several key steps:

1. Distillation

The first step in refining crude oil is distillation, which separates the oil into different fractions based on their boiling points. The crude oil is heated in a distillation column, causing it to vaporize. As the vapor rises, it cools and condenses at different heights in the column, separating the oil into various products, such as:

- **Gases:** Like propane and butane.
- **Light Distillates:** Such as gasoline.
- **Kerosene:** Used in jet fuel.
- **Heavy Distillates:** Such as diesel and heating oil.
- **Residuum:** The heavy residue that can be further processed into asphalt and other products.

2. Cracking

After distillation, heavier fractions may undergo cracking to break down large hydrocarbon molecules into smaller, more valuable ones. There are two main types of cracking:

- **Thermal Cracking:** Uses high temperatures and pressures to break down heavy oil fractions.
- **Catalytic Cracking:** Uses a catalyst to facilitate the cracking process at lower temperatures, yielding higher quantities of gasoline and diesel.

3. Reforming

Reforming is a process that converts low-quality naphtha into high-octane gasoline. It involves rearranging the molecular structure of hydrocarbons to improve their performance. This process can also produce valuable by-products, such as hydrogen, which can be used in other refining processes.

4. Treatment and Blending

After cracking and reforming, the refined products may contain impurities that need to be removed. This is achieved through various treatment processes, including:

- **Desulfurization:** Removes sulfur compounds to produce cleaner fuels that comply with environmental regulations.
- **Hydrotreating:** Treats the oil with hydrogen to remove impurities and enhance the quality of the products.

Finally, different refined products are blended to meet specific market requirements, such as the octane rating of gasoline or the viscosity of diesel fuel.

Environmental Considerations

While oil extraction and refining are essential for meeting global energy demands, they also pose significant environmental challenges. Oil spills, habitat destruction, and greenhouse gas emissions are critical issues that arise from these processes.

As a response, the industry has been investing in cleaner technologies and stricter regulations to mitigate the environmental impact of oil extraction and refining. This includes improving spill response protocols, adopting more efficient drilling techniques, and investing in renewable energy alternatives.

Conclusion

The extraction and refining of oil is a complex journey that transforms crude oil into the fuels and products that power our lives. From exploration and drilling to refining and distribution, each step involves intricate processes and advanced technologies. As we navigate the future of energy, understanding how oil is extracted and refined is crucial not only for appreciating its role in our daily lives but also for making informed decisions about our energy consumption and environmental impact.

How Are Aircraft Designed and Tested?

The design and testing of aircraft is a complex and highly regulated process that blends engineering, physics, and innovative technology. From commercial airliners to military jets, the aviation industry relies on meticulous planning and rigorous testing to ensure that each aircraft is safe, efficient, and capable of performing its intended functions. This exploration of aircraft design and testing will delve into the multifaceted stages involved, showcasing the collaboration of various disciplines and the critical importance of safety and performance.

1. Conceptual Design

The journey of any aircraft begins with conceptual design, where engineers and designers brainstorm ideas based on market needs, technology trends, and regulatory requirements. During this phase, several factors are considered, including:

- **Purpose of the Aircraft:** Is it a passenger plane, cargo aircraft, or military fighter jet? Each type has unique design requirements.

- **Market Analysis:** Understanding the demand for specific aircraft helps shape design decisions, including size, range, and capacity.
- **Regulatory Standards:** Aircraft must comply with stringent regulations set by authorities such as the Federal Aviation Administration (FAA) or the European Union Aviation Safety Agency (EASA). These regulations dictate safety standards, noise levels, and environmental considerations.

2. Preliminary Design

Once the concept is finalized, the preliminary design phase begins. Engineers use computer-aided design (CAD) software to create detailed drawings and models of the aircraft. Key activities during this phase include:

- **Aerodynamics:** Engineers conduct aerodynamic analyses to study how air flows around the aircraft. This is crucial for determining the aircraft's lift, drag, and stability. Wind tunnel testing of scale models may be used to validate theoretical predictions.
- **Structural Design:** The airframe must be designed to withstand various loads and stresses during flight. Engineers analyze materials and structures to ensure they are lightweight yet strong enough for safety.

- **Systems Integration:** Modern aircraft are equipped with complex systems, including avionics, propulsion, and control systems. Designers must integrate these systems into the aircraft's overall architecture, ensuring compatibility and efficiency.

3. Detailed Design

In this phase, engineers refine their designs and prepare for manufacturing. This involves creating comprehensive specifications for every component of the aircraft, including:

- **Materials Selection:** Engineers choose materials based on weight, strength, cost, and environmental resistance. Common materials include aluminum, titanium, and composite materials.
- **Manufacturing Processes:** Detailed plans for how each component will be produced are developed. This may involve machining, molding, or additive manufacturing techniques.
- **Final Drawings and Models:** Finalized CAD models and technical drawings are created, serving as the blueprint for manufacturing.

4. Prototyping

Once the detailed design is complete, a prototype is built. This step is crucial for testing the aircraft's design in a real-world environment. Key activities during this phase include:

- **Construction of the Prototype:** A full-scale prototype is built, incorporating all the design features and systems. This prototype is often constructed using the same materials and techniques planned for the final production model.
- **Ground Testing:** Before taking to the skies, the prototype undergoes extensive ground testing. This includes static tests to measure structural integrity, as well as system checks for avionics, engines, and other critical systems.

5. Flight Testing

The flight testing phase is one of the most critical stages in aircraft development. It involves multiple test flights to evaluate the aircraft's performance and safety. Key aspects of flight testing include:

- **Test Flight Plan:** Engineers develop a detailed flight test plan that outlines the objectives of each flight, including maneuvers to be performed and data to be collected.
- **Data Collection:** During flight tests, engineers gather extensive data on the aircraft's performance, including speed, altitude, fuel efficiency, and handling characteristics. Instrumentation is installed

on the aircraft to monitor various parameters in real time.
- **Pilot Evaluation:** Experienced test pilots conduct the flights, providing feedback on handling, comfort, and overall performance. Their insights are invaluable in identifying potential issues and refining the aircraft's design.

6. Certification

After successful flight testing, the aircraft must undergo a certification process. This step ensures that the aircraft meets all regulatory requirements and is safe for public use. Key components of the certification process include:

- **Documentation Submission:** Manufacturers must submit extensive documentation to regulatory authorities, detailing the design, testing results, and compliance with safety standards.
- **Third-Party Review:** Independent experts may review the aircraft's design and test data to ensure it meets safety and performance standards.
- **Certification Flight Tests:** Regulatory authorities may conduct their own flight tests to validate the aircraft's performance and safety before granting certification.

7. Production and Delivery

Once the aircraft is certified, it moves into production. This phase includes:

- **Manufacturing:** The aircraft is produced in a facility designed for high-quality, large-scale production. Assembly lines are set up to ensure efficiency and consistency.
- **Final Testing:** Each aircraft undergoes final testing before delivery to customers. This includes verifying that all systems function as intended and that the aircraft meets quality standards.
- **Delivery and Support:** After passing final inspections, the aircraft is delivered to the customer, whether it be an airline, military branch, or private owner. Manufacturers often provide ongoing support and maintenance services to ensure the aircraft remains operational throughout its lifespan.

Conclusion

The design and testing of aircraft are intricate processes that combine creativity, engineering expertise, and rigorous safety protocols. Each stage, from conceptual design to production, involves collaboration among various disciplines and stakeholders, ensuring that every aircraft is not only functional but also safe for those who fly in it. As technology continues to evolve, so too will the methods used to design and test aircraft, paving the way

Behind the Scenes

for the next generation of aviation innovations. Whether you're an aviation enthusiast or simply curious about how things are made, understanding this process reveals the dedication and ingenuity that go into every flight.

Behind the Scenes

How is the Internet Connected Globally?

In our hyper-connected world, the internet is an integral part of daily life. Whether we're streaming our favorite shows, conducting business, or connecting with loved ones across the globe, the internet serves as a vital backbone for communication and information exchange. But how is this vast network of networks established and maintained? This exploration delves into the intricate systems and technologies that connect the internet globally, highlighting the physical infrastructure, protocols, and the key players involved in creating this digital realm.

1. The Physical Infrastructure

At the heart of global internet connectivity lies a complex physical infrastructure consisting of cables, data centers, and satellites. Here are the key components:

a. Undersea Cables

A significant portion of international internet traffic is transmitted through undersea fiber-optic cables. These cables, often thousands of miles long, run along the ocean floor and connect continents.

Behind the Scenes

- **Construction and Installation:** Laying these cables is a monumental engineering feat. Specialized ships are used to deploy the cables, bury them to protect against external damage, and maintain the systems over time.
- **Capacity and Speed:** Fiber-optic cables transmit data as pulses of light, enabling high-speed communication over vast distances. Modern cables can carry terabits of data per second, far surpassing previous technologies.

b. Data Centers

Data centers are facilities that house the servers, storage systems, and networking equipment essential for processing and storing data. They are strategically located to optimize connectivity and redundancy.

- **Geographical Distribution:** Data centers are spread across the globe, allowing for localized processing and efficient data delivery. This distribution minimizes latency and ensures users have quick access to the information they need.
- **Interconnection:** Data centers are interconnected through high-capacity fiber-optic links, forming the backbone of the internet. They exchange data and provide redundancy in case of outages, enhancing overall reliability.

Behind the Scenes

c. Satellite Networks

For remote areas where laying cables is impractical, satellite networks provide an alternative means of internet connectivity.

- **Geostationary Satellites:** Positioned at fixed points above the Earth, these satellites allow for consistent coverage over specific regions. However, they may experience higher latency due to the distance the signals must travel.
- **Low Earth Orbit (LEO) Satellites:** Newer satellite constellations, like Starlink, utilize LEO satellites to provide faster and more reliable internet access, particularly in underserved areas. Their proximity to the Earth reduces latency, making them a promising solution for global connectivity.

2. Internet Protocols and Standards

Once the physical infrastructure is in place, a set of protocols governs how data is transmitted and received across networks. Key protocols include:

a. Transmission Control Protocol/Internet Protocol (TCP/IP)

TCP/IP is the foundational suite of communication protocols that facilitate data exchange on the internet.

Behind the Scenes

- **TCP:** This protocol ensures reliable transmission of data packets. It breaks data into smaller packets, sends them across the network, and reassembles them at the destination. If packets are lost or corrupted, TCP requests their retransmission.
- **IP:** The Internet Protocol is responsible for addressing and routing packets of data. Each device connected to the internet is assigned a unique IP address, allowing it to send and receive information.

b. Domain Name System (DNS)

The DNS is like the internet's phonebook, translating human-readable domain names (like www.example.com) into IP addresses that computers use to identify each other.

- **Hierarchical Structure:** The DNS operates through a hierarchical system of servers that work together to resolve domain names. This structure ensures fast and efficient lookups, enabling users to access websites quickly.
- **Caching:** To enhance performance, DNS servers cache recent lookups. This reduces the time it takes to resolve frequently accessed domains and decreases overall network traffic.

3. Internet Service Providers (ISPs)

Behind the Scenes

ISPs play a crucial role in connecting users to the internet. They provide the necessary infrastructure, support, and services for individuals and businesses.

a. Types of ISPs

- **Tier 1 ISPs:** These are major telecommunications companies that own extensive networks and can connect directly to the global internet backbone. They do not pay for internet transit; instead, they exchange traffic with other Tier 1 ISPs.
- **Tier 2 and Tier 3 ISPs:** These ISPs purchase access to the internet from Tier 1 ISPs and provide services to end-users, including residential customers and businesses. They may also operate their own local networks, extending connectivity to more remote areas.

b. Peering Agreements

ISPs engage in peering agreements to exchange traffic without incurring costs. These agreements enhance connectivity and enable data to flow efficiently between different networks.

- **Public Peering:** ISPs connect at Internet Exchange Points (IXPs), allowing multiple networks to interconnect and exchange traffic in a neutral environment.

- **Private Peering:** Two ISPs establish a direct connection to exchange traffic, enhancing speed and reliability.

4. The Role of Government and Regulation

Governments and regulatory bodies play an essential role in shaping the internet landscape. They establish rules and regulations that govern internet access, data privacy, and cybersecurity.

a. Policy Development

- Net Neutrality: The principle of net neutrality advocates that all internet traffic should be treated equally, without preferential treatment for specific websites or services. Regulatory debates around this principle impact how ISPs operate and affect user access.
- **International Cooperation:** As the internet transcends national borders, international agreements and collaborations are vital for addressing challenges such as cybersecurity, data privacy, and digital rights.

5. The Future of Global Connectivity

As technology continues to evolve, the quest for global connectivity remains ongoing. Innovations in infrastructure, such as advanced fiber-optic technologies and satellite systems, promise to improve internet access in underserved areas.

a. 5G and Beyond

The rollout of 5G networks offers faster speeds and lower latency, enabling new applications like smart cities and the Internet of Things (IoT). As 5G technology matures, it will enhance global connectivity and provide new opportunities for innovation.

b. The Digital Divide

Efforts to bridge the digital divide between urban and rural areas, as well as between developed and developing nations, are critical for ensuring equitable access to the internet. Organizations and governments are working to expand broadband access to underserved populations, empowering individuals with knowledge and resources.

Conclusion

Understanding how the internet is connected globally reveals a complex interplay of physical infrastructure, protocols, and the collaborative efforts of various stakeholders. As the internet continues to evolve, so too will the methods and technologies that connect us. By grasping the intricacies of this global network, we can

Behind the Scenes

appreciate the profound impact the internet has on our lives and the world at large.

Behind the Scenes

How is the Human Genome Mapped?

Mapping the human genome is one of the most significant scientific endeavors of the past few decades, leading to groundbreaking advancements in genetics, medicine, and our understanding of human biology. This complex process involves identifying and recording the locations of genes and other important markers within our DNA. Here, we will explore the steps, technologies, and implications of human genome mapping.

The Human Genome Project

The quest to map the human genome began in earnest with the launch of the Human Genome Project (HGP) in 1990, an international collaboration aimed at sequencing the entire human genome. The HGP sought to identify all the genes present in human DNA and determine their sequences. The project was completed in 2003, marking a monumental milestone in the field of genetics.

Steps in Mapping the Human Genome

1. **Sample Collection:** The first step in mapping the genome involves collecting DNA samples. These samples can be obtained from a variety of sources,

Behind the Scenes

including blood, saliva, or other tissues from volunteers. In the early days of the HGP, samples primarily came from anonymous donors to ensure privacy.

2. **DNA Sequencing:** Once DNA samples are collected, the next step is sequencing. Sequencing involves determining the exact order of the nucleotide bases (adenine, thymine, cytosine, and guanine) that make up the DNA. There are several methods of DNA sequencing, with the most significant being the Sanger sequencing method and next-generation sequencing (NGS).

    - **Sanger Sequencing:** Developed in the late 1970s, this method uses chain-terminating inhibitors to produce fragments of different lengths, which are then separated by size using gel electrophoresis. Sanger sequencing was instrumental during the initial phases of the HGP.
    - **Next-Generation Sequencing (NGS):** This newer approach allows for much faster and cheaper sequencing of large amounts of DNA. NGS technology involves massively parallel sequencing, where millions of fragments of DNA are sequenced simultaneously. This method has significantly accelerated the pace of genome mapping.

3. **Bioinformatics:** Mapping the genome generates an immense amount of data that needs to be analyzed and interpreted. Bioinformatics, a field that combines biology, computer science, and statistics, plays a crucial role in managing this data. Computational tools are used to align sequenced DNA fragments, identify genes, and annotate the genome.
4. **Genome Assembly:** The next step is assembling the sequenced DNA fragments into a complete genome. This process involves aligning overlapping fragments to reconstruct the original sequence. Various algorithms and software tools assist in this assembly, which can be a complex task due to the repetitive nature of certain regions in the genome.
5. **Annotation:** Once the genome is assembled, scientists annotate it to identify the locations and functions of genes, regulatory elements, and other significant features. This annotation process is essential for understanding the biological significance of various genomic regions and how they contribute to human health and disease.

Advances Beyond the HGP

Following the completion of the Human Genome Project, researchers continued to refine genome mapping techniques. Advances in technology, such as long-read sequencing and single-cell sequencing, have allowed for a more comprehensive understanding of genetic variation and complexity.

1. **Long-Read Sequencing:** Traditional sequencing methods often struggle with repetitive regions in the genome. Long-read sequencing technologies can produce longer sequences, allowing for better assembly of these complex areas.
2. **Single-Cell Sequencing:** This technique enables scientists to analyze the genomes of individual cells, revealing insights into cellular diversity and the genetic basis of diseases like cancer.
3. **Genome Editing Technologies:** Tools such as CRISPR-Cas9 have emerged from genome mapping research, enabling precise editing of genetic sequences. This technology has far-reaching implications for treating genetic disorders and advancing biotechnology.

Implications of Mapping the Human Genome

Mapping the human genome has profound implications for medicine, genetics, and our understanding of human biology:

- **Personalized Medicine:** With the knowledge gained from the human genome, healthcare can be tailored to individual patients based on their genetic

makeup. This approach allows for more effective treatments and the potential to prevent diseases before they manifest.
- **Understanding Genetic Diseases:** Genome mapping has enhanced our understanding of the genetic basis of various diseases, leading to the identification of mutations associated with conditions like cystic fibrosis, sickle cell anemia, and many forms of cancer.
- **Evolutionary Insights:** The comparative analysis of human genomes with those of other species has provided valuable insights into human evolution, migration patterns, and the genetic factors that make us unique.
- **Ethical Considerations:** The mapping of the human genome raises ethical questions regarding genetic privacy, discrimination, and the potential for gene editing. As we navigate these issues, it is crucial to ensure that the benefits of this knowledge are balanced with respect for individual rights.

Conclusion

The mapping of the human genome represents one of the most remarkable achievements in modern science. Through a combination of advanced technologies, collaboration, and an unwavering pursuit of knowledge, researchers have unlocked secrets hidden within our DNA.

Behind the Scenes

The impact of this work is far-reaching, paving the way for new medical treatments, a deeper understanding of human biology, and the ongoing exploration of what it means to be human in a rapidly advancing world. As we continue to unravel the complexities of the human genome, the potential for discovery and innovation remains limitless.

Behind the Scenes

How Are Robots Designed and Built?

The world of robotics is an intricate blend of art, science, and engineering, where creativity meets technology to create machines that can perform tasks autonomously or with human guidance. From simple robotic vacuum cleaners to complex humanoid robots and industrial automation systems, the design and construction of robots involve several essential steps and considerations. This guide will explore the fascinating process of designing and building robots, highlighting the various disciplines involved and the technologies utilized.

1. Understanding the Purpose and Requirements

The first step in designing a robot is to clearly define its purpose. This involves identifying the specific tasks the robot will perform, the environment in which it will operate, and the desired level of autonomy. For example, a robot designed for manufacturing will have different requirements than one intended for home assistance or exploration in hostile environments. This stage often involves:

- **Research and Feasibility Analysis:** Understanding the problem the robot aims to solve, the market demand, and the existing solutions.
- **Specifications Development:** Defining the robot's capabilities, such as speed, load capacity, precision, and operational range.

2. Conceptual Design

Once the requirements are clear, engineers and designers move on to conceptualizing the robot. This involves brainstorming ideas and creating preliminary designs. Key activities include:

- **Sketching and Modeling:** Designers create sketches and use computer-aided design (CAD) software to develop 3D models of the robot. This helps visualize the robot's appearance, size, and configuration.
- **Selecting Components:** Choosing the appropriate sensors, actuators, processors, and other components based on the robot's intended functions. For example, a mobile robot might require wheels or tracks, while a robotic arm might need servos or motors.

3. Detailed Design and Prototyping

Behind the Scenes

In this phase, the conceptual design is transformed into a detailed blueprint. This involves:

- **Engineering Calculations:** Performing calculations to ensure that the robot can handle the physical demands of its tasks. This includes analyzing load distributions, torque requirements, and power needs.
- **Electronics and Software Development:** Designing the electronic circuits that control the robot's components and writing the software that governs its operations. This can include programming algorithms for navigation, object recognition, and task execution.
- **Prototyping:** Creating a physical prototype of the robot, often using rapid prototyping techniques such as 3D printing. This allows engineers to test and refine the design before mass production.

4. Building the Robot

Once the prototype has been validated, the next step is to construct the robot. This involves:

- **Assembling Components:** Carefully assembling all mechanical parts, electronic circuits, and software into the final robot. This may require precision tools and techniques to ensure everything fits and functions correctly.

- **Integrating Systems:** Ensuring that all components, such as sensors, motors, and controllers, communicate effectively. This often involves configuring wiring and ensuring proper signal flow.

5. Testing and Iteration

After assembly, thorough testing is crucial to ensure the robot performs as intended. This process involves:

- **Functional Testing:** Checking that all components work correctly and that the robot can perform its designated tasks.
- **Stress Testing:** Evaluating the robot's performance under extreme conditions or loads to ensure it can operate reliably in real-world scenarios.
- **Iteration:** Based on test results, engineers may need to iterate on the design, making adjustments to components, software, or configurations to improve performance and reliability.

6. Final Deployment and Maintenance

Once testing is complete and the robot meets all performance standards, it is ready for deployment. This phase includes:

- **User Training:** Educating users on how to operate and maintain the robot effectively. This is particularly important for robots intended for complex tasks in industrial settings.
- **Maintenance Planning:** Establishing a maintenance schedule to ensure the robot continues to operate efficiently. This can involve regular software updates, hardware inspections, and component replacements as necessary.

7. Future Developments

The field of robotics is continuously evolving, with advancements in artificial intelligence, machine learning, and materials science leading to smarter, more capable robots. Designers and engineers are now focusing on:

- **Autonomy:** Enhancing the robot's ability to operate independently in dynamic environments using advanced AI algorithms.
- **Collaboration:** Developing robots that can work alongside humans safely and efficiently, often referred to as collaborative robots (cobots).
- **Miniaturization:** Creating smaller robots that can navigate tight spaces, perform delicate tasks, or operate in hazardous environments.

Conclusion

Behind the Scenes

The design and construction of robots are multi-faceted processes that require collaboration across various disciplines, including mechanical engineering, electronics, computer science, and human factors engineering. By understanding how robots are designed and built, we can appreciate the complexity and ingenuity involved in creating machines that augment our capabilities and enhance our lives. As technology continues to advance, the potential for robotic applications is limitless, promising exciting developments in the future.

Behind the Scenes

How is Renewable Energy Harnessed?

As the world grapples with the urgent need to reduce greenhouse gas emissions and combat climate change, renewable energy has emerged as a crucial alternative to fossil fuels. Unlike conventional energy sources, which are finite and contribute to environmental degradation, renewable energy harnesses the natural processes of the Earth to produce clean, sustainable power. This chapter explores the various forms of renewable energy, how they are harnessed, and their potential to transform our energy landscape.

Understanding Renewable Energy

Renewable energy comes from resources that are naturally replenished. These include sunlight, wind, rain, tides, waves, and geothermal heat. The core idea is to capture and convert these abundant natural resources into usable energy forms, such as electricity or heat.

Types of Renewable Energy Sources

1. **Solar Energy**
    - **Harnessing Method**: Solar energy is captured through photovoltaic (PV) cells or

solar thermal systems. PV cells convert sunlight directly into electricity using semiconductor materials, while solar thermal systems use sunlight to heat a fluid, which can then generate steam to drive a turbine and produce electricity.
- **Applications**: Solar panels can be installed on rooftops, in solar farms, or even integrated into building materials. They provide electricity for residential, commercial, and industrial use and can also power devices directly.

2. **Wind Energy**
   - **Harnessing Method**: Wind energy is harnessed using wind turbines, which convert kinetic energy from wind into mechanical power. As wind blows across the blades of a turbine, it causes them to spin, driving a generator that produces electricity.
   - **Applications**: Wind farms can be found on land (onshore) or offshore, where winds are typically stronger and more consistent. Wind energy is used to power homes, businesses, and even entire cities.

3. **Hydropower**
   - **Harnessing Method**: Hydropower, or hydroelectric power, generates electricity by

using the flow of water. Dams are built on rivers to create reservoirs, and as water is released, it flows through turbines, generating electricity.
    - **Applications**: Hydropower is a significant source of renewable energy worldwide, providing electricity for millions of people. Small-scale hydroelectric systems can also be used in rural areas to provide localized energy solutions.
4. **Biomass Energy**
    - **Harnessing Method**: Biomass energy comes from organic materials, such as plant and animal waste. These materials can be burned directly for heat or converted into biofuels (like ethanol and biodiesel) through biological and chemical processes.
    - **Applications**: Biomass can be used for heating, electricity generation, and as a replacement for fossil fuels in transportation. Sustainable biomass practices involve using waste products and ensuring that sources are renewable.
5. **Geothermal Energy**
    - **Harnessing Method**: Geothermal energy utilizes heat from beneath the Earth's surface. Wells are drilled into geothermal reservoirs to

bring hot water or steam to the surface, which can then be used to drive turbines and generate electricity.
    - **Applications**: Geothermal energy is often used for direct heating (like district heating systems) and can provide reliable baseload power, making it a stable renewable energy source.
6. **Ocean Energy**
    - **Harnessing Method**: Ocean energy encompasses various technologies that capture energy from the sea, including tidal energy (the gravitational pull of the moon and sun), wave energy (the movement of ocean waves), and ocean thermal energy conversion (using temperature differences between surface and deep water).
    - **Applications**: While still in the early stages of development compared to other renewables, ocean energy has significant potential to contribute to energy needs, particularly in coastal regions.

The Process of Harnessing Renewable Energy

The process of harnessing renewable energy involves several key steps:

1. **Resource Assessment**: Identifying the available renewable resources in a specific area is the first step. This includes analyzing wind speeds, solar irradiance, water flow rates, and geothermal potential.
2. **Technology Selection**: Different technologies are chosen based on the resource type and local conditions. This could involve selecting solar panels, wind turbines, or geothermal heat pumps.
3. **Infrastructure Development**: Constructing the necessary infrastructure is crucial. For solar, this means installing panels; for wind, erecting turbines; for hydropower, building dams or channels.
4. **Energy Conversion**: Renewable resources must be converted into usable energy forms. This may involve generating electricity through turbines, heating fluids for thermal energy, or producing biofuels.
5. **Distribution**: Once energy is generated, it must be transmitted and distributed to consumers. This requires connecting renewable energy sources to the grid and ensuring efficient delivery.
6. **Integration and Storage**: Integrating renewable energy into existing energy systems can be complex, especially due to variability in generation (e.g., less solar power at night). Energy storage solutions, like

batteries, are essential for storing excess energy for use when production is low.

The Benefits of Renewable Energy

Harnessing renewable energy has numerous benefits:

- **Environmental Impact**: Renewable energy significantly reduces greenhouse gas emissions and air pollutants, mitigating climate change and improving air quality.
- **Energy Security**: By diversifying energy sources, countries can reduce dependence on imported fossil fuels and enhance energy security.
- **Job Creation**: The renewable energy sector has proven to be a significant source of job growth, from manufacturing to installation and maintenance.
- **Sustainable Development**: Renewable energy can provide reliable electricity to underserved and rural communities, supporting economic development and improving quality of life.

Challenges Ahead

Despite its advantages, renewable energy faces challenges, including:

- **Intermittency**: Solar and wind energy are variable sources, requiring advancements in energy storage and grid management.
- **Infrastructure Needs**: Developing the infrastructure to harness and distribute renewable energy can be costly and time-consuming.
- **Policy and Regulation**: Supportive policies and regulations are essential to encourage investment and facilitate the transition to renewable energy.

Conclusion

As the world moves towards a more sustainable future, harnessing renewable energy will play a pivotal role in achieving climate goals and ensuring a cleaner environment. By understanding how renewable energy is generated and integrated into our lives, we can make informed choices and support the transition towards a greener, more sustainable energy system. This journey toward a renewable future requires innovation, investment, and collective effort, and every step taken today brings us closer to a world powered by clean energy.

Behind the Scenes

How Are Satellites Launched into Orbit?

Satellites have become an integral part of our daily lives, enabling everything from GPS navigation to weather forecasting, telecommunications, and space exploration. However, the journey of a satellite from the drawing board to its designated orbit is a complex and meticulously planned process that involves several stages, advanced technology, and a significant amount of collaboration among various sectors. This chapter delves into the fascinating world of satellite launches, exploring how these remarkable machines are propelled into orbit around Earth.

Understanding Satellites and Their Orbits

Before diving into the launch process, it's essential to understand what satellites are and the types of orbits they occupy. A satellite is an artificial object placed into orbit around Earth or other celestial bodies. These satellites can serve various functions, such as communication, observation, navigation, and scientific research.

Types of Orbits

Behind the Scenes

1. **Low Earth Orbit (LEO):** Typically between 160 to 2,000 kilometers (99 to 1,242 miles) above Earth, satellites in LEO are used for Earth observation, imaging, and communication. Examples include the International Space Station (ISS) and many Earth observation satellites.
2. **Medium Earth Orbit (MEO):** Ranging from 2,000 to 35,786 kilometers (1,242 to 22,236 miles), this orbit is often used for navigation satellites like the Global Positioning System (GPS).
3. **Geostationary Orbit (GEO):** Positioned approximately 35,786 kilometers (22,236 miles) above the equator, satellites in GEO maintain a fixed position relative to the Earth's surface. This orbit is commonly used for communication and weather satellites, as it allows them to provide consistent coverage of specific areas.
4. **Polar Orbit:** Satellites in polar orbits pass over the Earth's poles, allowing them to cover the entire surface of the planet over time. This type of orbit is often used for Earth observation satellites.

The Launch Process

Launching a satellite into orbit is a multi-step process that involves extensive planning, engineering, and execution. Here's a breakdown of how it works:

1. Design and Development

The first step in launching a satellite is designing and developing it to meet specific mission requirements. This phase includes:

- **Mission Planning:** Engineers and scientists define the satellite's purpose, functionality, and operational requirements. This may involve assessing the type of orbit needed and the instruments required for the mission.
- **Satellite Design:** Engineers create detailed designs and specifications for the satellite, ensuring it can withstand the harsh conditions of space, including extreme temperatures, radiation, and vacuum.
- **Testing:** The satellite undergoes rigorous testing, including thermal vacuum tests, vibration tests, and electromagnetic compatibility tests, to ensure its reliability and functionality in space.

2. Manufacturing

Once the design is finalized, the satellite is manufactured. This process involves:

- **Component Assembly:** Various components, including the payload (the instruments that will perform the satellite's function), power systems,

communication systems, and propulsion systems, are assembled.
- **Integration:** The satellite is integrated into a single unit, ensuring that all systems work together seamlessly. This includes wiring, software installation, and system checks.

3. Launch Vehicle Selection

Choosing the right launch vehicle is crucial for a successful satellite launch. The launch vehicle, commonly known as a rocket, is responsible for transporting the satellite from the ground to its intended orbit. Considerations for vehicle selection include:

- **Payload Capacity:** The rocket must be capable of carrying the satellite's weight and dimensions.
- **Orbit Requirements:** Different rockets are designed for different types of orbits. For example, some rockets are optimized for LEO, while others are better suited for GEO.
- **Launch Provider:** Various companies, including SpaceX, United Launch Alliance, and Arianespace, provide launch services. The selection often depends on cost, reliability, and schedule availability.

4. Pre-Launch Preparations

Before the actual launch, several critical steps are taken:

- **Launch Site Preparation:** The launch site is readied, which includes securing the area, setting up infrastructure, and conducting final checks on the rocket and satellite.
- **Transportation to the Launch Pad:** The assembled rocket, with the satellite securely attached, is transported to the launch pad. This may involve specialized vehicles and careful handling to avoid damage.
- **Final Checks:** A series of final inspections and checks are performed to ensure that all systems are functioning correctly. This includes verifying fuel levels, communication systems, and telemetry.

5. The Launch

The launch itself is a highly synchronized and carefully orchestrated event. Key elements include:

- **Countdown:** A countdown sequence begins, during which final checks are performed, and the rocket is fueled.
- **Ignition and Lift-off:** The rocket engines are ignited, and the vehicle lifts off the launch pad. It experiences a tremendous amount of thrust, which propels it into the sky.
- **Ascent Phase:** The rocket ascends through the atmosphere, often experiencing several stages of

propulsion. Most rockets are multi-stage, meaning they shed sections of themselves as fuel is consumed to reduce weight.

6. Reaching Orbit

Once the rocket reaches its designated altitude, it goes through the following phases:

- **Staging:** In multi-stage rockets, the first stage will separate once its fuel is exhausted. The remaining stages continue the journey into orbit.
- **Payload Deployment:** Once the final stage of the rocket reaches the desired orbit, it releases the satellite into space. This is done with precision to ensure the satellite is on the correct trajectory.
- **Orbit Insertion:** After deployment, the satellite uses its onboard propulsion system to maneuver into its operational orbit. This may involve adjusting its altitude and orientation.

7. Post-Launch Operations

After the satellite is in orbit, several critical tasks are performed:

- **Initial Checks:** Engineers conduct a series of checks to ensure that all systems are functioning correctly after deployment.

- **Commissioning:** The satellite undergoes a commissioning phase where it is tested and calibrated to ensure it meets mission requirements. This may involve activating scientific instruments, establishing communication links, and adjusting orbits.
- **Operational Phase:** Once commissioning is complete, the satellite begins its operational phase, carrying out its designated tasks for the duration of its mission, which can last from a few years to several decades.

Conclusion

The launch of a satellite into orbit is a remarkable achievement that embodies human ingenuity, collaboration, and technological advancement. From the initial design phase to the complexities of the launch process, each step is critical to ensuring the success of the mission. As we continue to push the boundaries of space exploration and technology, understanding how satellites are launched into orbit provides valuable insight into the intricate processes that enable us to connect, communicate, and explore our universe. Whether it's for scientific discovery, navigation, or global communications, satellites will continue to play a pivotal role in shaping our world and expanding our understanding of the cosmos.

Behind the Scenes

How Is Paper Recycled?

Paper recycling is an essential process that helps reduce waste, save energy, and conserve natural resources. With increasing environmental awareness, recycling paper has become a vital part of waste management and sustainability practices worldwide. But how exactly is paper recycled? Let's take a closer look at the stages involved in recycling paper, from the collection of used paper to the production of new paper products.

1. Collection

The recycling process begins with the collection of used paper. This can occur at various locations, including homes, offices, and industrial sites. Many municipalities have designated recycling bins for paper waste, while some businesses implement their own recycling programs. After collection, the paper is transported to a recycling facility.

2. Sorting

Once at the recycling facility, the collected paper is sorted into different categories based on type, grade, and quality. This is a crucial step, as different types of paper require different recycling processes. For instance, office paper, newspaper, cardboard, and magazines each have unique characteristics. Sorting helps ensure that the recycling process is efficient and that the resulting recycled paper products meet the required standards.

3. Shredding

After sorting, the paper is shredded into small pieces. This shredding process makes it easier to break down the paper fibers during the subsequent stages of recycling. It also helps remove contaminants, such as plastic, metal, or other materials that may have been mixed in with the paper waste.

4. Pulping

The shredded paper is then mixed with water and chemicals to create a slurry known as pulp. This mixture is agitated to separate the paper fibers from each other. The pulping process can be done using two main methods:

- **Mechanical Pulping:** In this method, paper is physically ground down into pulp using machines. While this method is energy-intensive, it preserves more of the paper fibers, making it suitable for producing certain types of recycled paper products.
- **Chemical Pulping:** This process involves using chemicals, such as sodium hydroxide and sodium sulfide, to break down the paper fibers. Chemical pulping results in higher-quality pulp with fewer

Behind the Scenes

impurities, making it ideal for producing high-quality paper products.

5. Cleaning

Once the pulp is formed, it undergoes a cleaning process to remove any remaining contaminants, such as ink, adhesives, and non-paper materials. This can involve several methods, including flotation, where air bubbles are introduced to the pulp to lift contaminants to the surface, or screening, which separates larger particles from the pulp.

6. Bleaching (Optional)

If the recycled paper is to be used for producing white or high-quality paper products, a bleaching process may be employed. Bleaching involves using chemicals to lighten the color of the pulp and remove any remaining dyes. However, many recycling facilities are now moving toward environmentally friendly bleaching methods to minimize the use of harmful chemicals.

7. Papermaking

Once the pulp is cleaned and bleached (if necessary), it is ready for the papermaking process. The pulp is diluted with water and then poured onto a flat screen that allows the water to drain away. This forms a wet sheet of paper.

The sheet is then pressed to remove excess water and dried using heated rollers.

8. Finishing and Cutting

After the paper has dried, it undergoes finishing processes, which may include calendering (smoothing the paper) and coating (applying a surface layer for gloss or durability). Finally, the finished paper is rolled up and cut into sheets, ready for packaging and distribution.

9. Reuse

The recycled paper can now be used to produce a wide range of products, including:

- Writing paper
- Cardboard boxes
- Tissue paper
- Newsprint
- Paperboard for packaging

Benefits of Paper Recycling

Recycling paper offers numerous environmental and economic benefits:

- **Conservation of Resources:** Recycling paper reduces the need for virgin wood pulp, which helps preserve forests and natural habitats.

- **Energy Savings:** The recycling process generally requires less energy compared to producing paper from raw materials. Studies suggest that recycling paper can save up to 60% of the energy used in paper production.
- **Reduction of Landfill Waste:** Recycling paper decreases the volume of waste sent to landfills, reducing methane emissions and extending the lifespan of existing landfills.
- **Lower Carbon Footprint:** By conserving resources and reducing energy consumption, recycling paper helps lower greenhouse gas emissions associated with paper production.

Conclusion

The process of recycling paper is a vital aspect of environmental sustainability. From collection to processing, every step in the recycling chain plays a crucial role in transforming used paper into new products. By understanding how paper is recycled, we can appreciate the importance of this practice in conserving natural resources and protecting our planet.

So the next time you toss a piece of paper into the recycling bin, remember that you're contributing to a cycle that not only helps reduce waste but also supports a more sustainable future for generations to come.

Behind the Scenes

How Are Fireworks Made?

Fireworks have dazzled audiences for centuries, filling the sky with vibrant colors, dazzling patterns, and the thrill of loud explosions. From celebratory events like New Year's Eve and Independence Day to cultural festivals around the world, fireworks create unforgettable experiences. But have you ever wondered how these spectacular displays are made? The process of crafting fireworks is a blend of art and science, involving precise engineering, chemistry, and creativity. Let's delve into the fascinating world of fireworks manufacturing to understand how these aerial wonders come to life.

The Components of Fireworks

At the heart of every firework are its essential components, each playing a vital role in creating the final display. Understanding these components is the first step in grasping how fireworks are made:

1. **Shell:** The shell is the outer casing that holds the various components together. Typically made from a paper or cardboard tube, it is designed to withstand the pressure of the explosion and protect the contents until they are ignited.

281

2. **Pyrotechnic Stars:** These are the colorful particles that create the dazzling displays. They are made from a mixture of oxidizers, fuel, and metal salts that produce different colors when ignited. The choice of metals used determines the color of the fireworks: for example, strontium produces red, barium yields green, and sodium creates yellow.
3. **Black Powder (Gunpowder):** Black powder is the primary propellant in fireworks. It is made from a mixture of potassium nitrate, charcoal, and sulfur. When ignited, it creates a rapid expansion of gas that propels the firework into the air and ignites the stars inside.
4. **Fuses:** Fuses are essential for igniting the fireworks. They come in various types, including time-delay fuses that allow for the timed ignition of multiple shells during a display.
5. **Additional Effects:** Many fireworks include additional effects, such as crackles, whistles, or smoke, which are achieved by adding specific chemicals or materials to the mixture.

The Manufacturing Process

The production of fireworks is a meticulous process that combines artistry with strict safety measures. Here's a step-by-step overview of how fireworks are made:

Behind the Scenes

1. Ingredient Preparation

The first step in making fireworks is gathering and preparing the raw materials. This involves sourcing high-quality chemicals, including oxidizers, fuels, and metal salts. The materials are often ground into fine powders to ensure they mix evenly.

2. Mixing the Chemical Compositions

Once the ingredients are ready, they are mixed in specific proportions to create the desired pyrotechnic stars and effects. This step is crucial, as the correct ratios determine the colors and effects produced when the firework is ignited. Special care is taken to prevent accidental ignition during this process, as many of the materials are highly flammable.

3. Creating the Stars

The mixed compositions are then pressed into small pellets or stars, typically about the size of a marble. These stars will be packed into the shell later. The stars are sometimes coated with additional layers to enhance their color and brightness.

4. Assembling the Shells

The shells are constructed from paper or cardboard tubes, which are often reinforced for durability. The shells are designed with specific shapes and sizes, depending on the type of firework being made. Once the shells are ready, they are packed with black powder and the pyrotechnic stars.

5. Adding the Fuses

Fuses are inserted into the shells at this stage. Depending on the design, the fuses may connect to multiple shells for a synchronized display. Each fuse is carefully tested to ensure it will ignite the black powder reliably.

6. Final Assembly and Packaging

Once all components are assembled, the shells are sealed and packaged for distribution. Each firework undergoes quality control checks to ensure that it meets safety standards and will perform as intended.

Safety Precautions

Given the volatile nature of fireworks, safety is paramount throughout the manufacturing process. Factories are equipped with specialized ventilation systems, fire-resistant materials, and strict protocols to minimize the risk of accidents. Workers are trained to handle materials safely, and numerous regulations govern the manufacturing and transportation of fireworks.

The Art of Pyrotechnics

While the manufacturing of fireworks is a science, it is also an art. Skilled pyrotechnicians design and choreograph firework displays to create a breathtaking experience. They consider timing, color combinations, and effects to synchronize the explosions with music or a particular theme. This artistry transforms the raw materials into a stunning visual spectacle.

Conclusion

Fireworks are a remarkable blend of science, engineering, and art. The intricate process of creating these dazzling displays involves careful preparation, skilled craftsmanship, and a deep understanding of chemistry. The next time you watch a firework light up the night sky, you'll have a newfound appreciation for the effort and expertise that goes into making each spectacular explosion possible. From the initial ingredient preparation to the final launch, the journey of fireworks from concept to celebration is a testament to human creativity and ingenuity.

Behind the Scenes

How Are Pandemics Controlled?

Pandemics, by their very nature, spread rapidly across continents, impacting vast populations and overwhelming health systems. The challenge of controlling a pandemic requires a coordinated global response that combines public health measures, scientific advancements, government interventions, and social cooperation. From historical outbreaks like the Black Death to modern-day challenges such as COVID-19, the approach to controlling pandemics has evolved significantly. So, how are pandemics controlled today?

1. Early Detection and Surveillance

The first line of defense in controlling a pandemic is early detection. Health organizations, such as the World Health Organization (WHO) and the Centers for Disease Control and Prevention (CDC), continuously monitor disease outbreaks around the world. These systems rely on real-time data from hospitals, clinics, and research centers to identify unusual patterns or clusters of illnesses that could signal a new outbreak.

With advancements in technology, surveillance systems are now more sophisticated, employing tools such as

artificial intelligence to analyze data and detect potential pandemics more rapidly. Genetic sequencing also plays a crucial role in identifying new viruses and tracking their mutations, allowing for faster and more accurate responses.

2. Quarantine and Isolation Measures

Once an outbreak is detected, one of the most immediate steps taken to control it is **quarantine** and **isolation**. Quarantine restricts the movement of individuals who may have been exposed to the disease but are not yet symptomatic, while isolation is used for individuals who are already sick to prevent the spread to others.

These measures were historically implemented during pandemics like the 1918 Spanish flu and remain essential in modern outbreaks. However, with the advent of air travel and global interconnectivity, enforcing quarantines and isolating individuals quickly enough to stop the spread requires precise planning and international cooperation.

3. Contact Tracing

Contact tracing is another vital tool in controlling pandemics. Once an infected person is identified, public health officials trace their recent contacts to locate others who may have been exposed to the virus. This helps break the chain of transmission by identifying individuals at risk and taking appropriate measures, such as testing, quarantining, or isolating them.

Contact tracing has been revolutionized by mobile technology. Apps and GPS-based tools now enable authorities to track potential exposure more efficiently, though these methods raise concerns about privacy and data security.

4. Vaccination Campaigns

Vaccines are one of the most powerful tools in controlling pandemics. Historically, the development of vaccines for diseases like smallpox and polio has eradicated or greatly reduced their spread. In the case of a novel virus, however, vaccines need to be developed from scratch. This requires **genetic analysis** of the virus, clinical trials to ensure safety and efficacy, and finally, mass production and distribution.

In the case of COVID-19, scientists achieved unprecedented speed in developing vaccines, some within a year of the outbreak, largely thanks to advances in **mRNA technology** and global collaboration. Mass vaccination campaigns aim to achieve **herd immunity**, where a large enough portion of the population is immune to the virus, reducing its ability to spread.

5. Public Health Communication

Clear, accurate, and timely communication is key to controlling pandemics. Public health authorities must inform the public about the severity of the situation, how to protect themselves, and what measures are being taken. Effective communication helps build trust and encourages cooperation with health guidelines, such as mask-wearing, social distancing, and hygiene practices.

Misinformation can significantly hinder efforts to control a pandemic, so combating myths and false information is just as important as spreading factual guidance. In recent pandemics, social media platforms have played a crucial role in both spreading information and misinformation, prompting health authorities to work closely with these platforms to ensure the public receives accurate updates.

6. Pharmaceutical Interventions

In addition to vaccines, **antiviral drugs** and **therapeutics** play an important role in treating infected individuals and reducing the severity of symptoms. During pandemics, pharmaceutical companies race to develop or repurpose existing drugs that can help alleviate the disease. For example, during the COVID-19 pandemic, drugs like remdesivir and dexamethasone were used to treat severe cases.

Monoclonal antibodies and other immune-based treatments are also deployed to help the body fight the virus, especially in high-risk patients. The faster these treatments can be developed, tested, and distributed, the better the chances of reducing the virus's impact.

7. Global Collaboration and Governance

Pandemics know no borders, and controlling them requires **international cooperation**. Organizations like the WHO coordinate global efforts by providing guidelines, funding research, and distributing medical supplies. Governments must also work together to share data, resources, and vaccines to ensure that the pandemic is addressed globally, not just within individual nations.

The COVID-19 pandemic demonstrated the importance of global solidarity. Initiatives like COVAX, designed to ensure equitable access to vaccines for low- and middle-income countries, showed how global collaboration could mitigate the impacts of a pandemic worldwide. However, challenges such as vaccine nationalism, where countries prioritize their own populations at the expense of others, highlight the need for more equitable solutions in future pandemics.

8. Non-Pharmaceutical Interventions (NPIs)

When vaccines or treatments are not yet available, governments and health organizations rely on **non-**

pharmaceutical interventions (NPIs) to slow the spread of disease. These include:

- **Social distancing**: Limiting close contact between individuals to prevent transmission.
- **Face masks**: Encouraging or mandating the use of masks to reduce the spread of respiratory droplets.
- **School and business closures**: Temporarily shutting down institutions to limit large gatherings.
- **Travel restrictions**: Imposing limits on domestic and international travel to contain the virus in certain areas.

The effectiveness of these measures depends on public compliance and swift governmental action. While NPIs can be highly effective, they also come with social and economic costs, making it essential for governments to carefully balance public health with economic stability.

9. Economic Support and Recovery Plans

Controlling a pandemic also involves mitigating its **economic impact**. Lockdowns, travel bans, and restrictions often lead to job losses, business closures, and economic recessions. Governments must implement economic support measures, such as **stimulus packages**, unemployment benefits, and small business loans, to ensure the survival of affected communities and businesses.

Effective pandemic control is not just about stopping the virus; it's about ensuring society can recover and rebuild afterward. Long-term economic recovery plans, infrastructure investments, and international aid are vital to help countries bounce back after the crisis has been managed.

Conclusion

Pandemics are global crises that require multi-faceted strategies to control. From early detection and surveillance to the deployment of vaccines and therapeutics, the fight against pandemics involves collaboration across all sectors of society—governments, health organizations, scientists, businesses, and the public. As our understanding of viruses grows and technology advances, so too does our ability to respond to and control pandemics more effectively. However, it is essential to remember that human behavior plays a crucial role in how quickly and successfully we can control these outbreaks.

The more we learn from past pandemics, the better equipped we are to handle the next one.

Behind the Scenes

How is a Supercomputer Built?

Supercomputers, the giants of the computing world, are essential for solving some of the most complex problems in science, medicine, engineering, and even weather forecasting. These machines are vastly more powerful than ordinary personal computers, capable of performing billions or even trillions of calculations per second. But how is a supercomputer built? In this chapter, we'll explore the intricate process of constructing one of the most powerful tools in modern technology, examining the hardware, architecture, and cooling systems required to keep these machines running at full speed.

1. The Purpose and Design

Before the physical building of a supercomputer can begin, it's essential to define the purpose for which the machine will be used. Supercomputers are typically designed to handle highly specialized tasks such as simulating weather patterns, running complex simulations of nuclear reactions, or analyzing massive datasets in fields like genomics and astrophysics. Understanding the machine's specific use cases helps engineers decide how to balance processing power, memory, and storage.

Behind the Scenes

Once the purpose is defined, engineers design the overall architecture of the supercomputer. This involves choosing the right balance between **processing units (CPUs and GPUs), memory, and storage systems**. These choices are critical, as different applications require different architectures. For instance, supercomputers used for simulations might need more powerful CPUs, while those used for artificial intelligence might rely heavily on GPUs for rapid parallel processing.

2. The Building Blocks: Processors and Nodes

At the heart of every supercomputer is its **processor**, often consisting of thousands or even millions of cores. Unlike regular computers, which may have 2 to 16 cores, supercomputers are built using **multi-core processors** to perform multiple tasks simultaneously. These processors are typically top-of-the-line CPUs or GPUs from companies like Intel, AMD, or Nvidia. Some supercomputers also use custom-designed processors, optimized for specific workloads.

These processors are arranged in units called **nodes**. A node is essentially a high-performance computer on its own, with its own processors, memory, and sometimes storage. A supercomputer may have thousands or tens of thousands of nodes, all connected together to work as one colossal computing system.

Behind the Scenes

The nodes are then linked together using a high-speed **interconnect** network, which allows them to communicate with each other. This network needs to be incredibly fast, as a bottleneck in communication would severely limit the supercomputer's overall performance. Technologies such as InfiniBand or proprietary interconnects are often used to ensure the fastest possible data transfer between nodes.

3. Memory and Storage

For a supercomputer to perform at its best, the right combination of **memory (RAM)** and **storage** is crucial. Supercomputers need vast amounts of memory to hold and process large datasets. Modern supercomputers are often equipped with terabytes or even petabytes of RAM. However, there is a trade-off: more memory means higher costs and increased power consumption.

Storage is another critical component. While RAM is used for short-term memory, **long-term storage** systems are necessary for saving and retrieving the enormous amounts of data that supercomputers generate and analyze. Supercomputers often use high-speed solid-state drives (SSDs) for rapid data access, but they may also include traditional hard drives for mass data storage. In some cases, cutting-edge storage systems like **distributed file systems** are used to ensure that data can be accessed quickly by thousands of nodes simultaneously.

Behind the Scenes

4. Cooling Systems

One of the most challenging aspects of building a supercomputer is dealing with the **heat** it generates. With thousands of processors running at full capacity, the machine can easily overheat, causing severe damage to its components. As a result, **cooling** becomes a top priority in the design of a supercomputer.

Supercomputers typically use advanced cooling technologies, such as **liquid cooling systems**, where coolant circulates around the processors and other key components to keep them at optimal temperatures. In some cases, supercomputers are submerged in special non-conductive liquids that can absorb heat more effectively than air. Other systems use traditional **air cooling** with industrial-scale fans, though these systems are less efficient for supercomputers with high power densities.

More innovative supercomputers, such as the ones built by companies like IBM and Cray, use sophisticated **direct liquid cooling**, where the cooling liquid is pumped directly through the chips themselves, ensuring the most efficient heat dissipation. These cooling systems often require large-scale **cooling towers** or even underground systems to manage the heat generated.

5. Power Supply and Energy Efficiency

Behind the Scenes

Running a supercomputer requires an immense amount of power. Some of the most powerful supercomputers consume as much electricity as a small town, making **power supply** and **energy efficiency** crucial design considerations.

Many modern supercomputers are built with energy efficiency in mind, using advanced power management technologies to minimize waste and ensure that the machine is as environmentally friendly as possible. **Green computing** initiatives are becoming increasingly important in the field, with many supercomputers now using renewable energy sources or incorporating power-saving designs that reduce their carbon footprint.

The **TaihuLight** supercomputer in China, for example, is renowned not only for its performance but also for its energy-efficient architecture, which allows it to achieve tremendous computational power without excessive energy consumption.

6. Software and Programming

The hardware is only one part of a supercomputer. For the system to operate effectively, specialized **software** is required. Supercomputers typically run on custom versions of Linux, and the software stack is designed to take full advantage of the system's massive parallelism.

Behind the Scenes

Parallel computing is the cornerstone of supercomputer operation. Programs need to be specifically designed to split tasks across thousands of processors, allowing the supercomputer to solve large-scale problems much faster than a regular computer could. Software development for supercomputers involves creating algorithms that can efficiently distribute the workload and minimize communication delays between nodes.

In addition, **job schedulers** are used to manage how different tasks are assigned to the processors. Supercomputers typically run many jobs simultaneously, and the scheduler ensures that each job is given the resources it needs while optimizing the overall performance of the machine.

7. Testing and Optimization

Once the hardware and software are in place, the supercomputer undergoes rigorous **testing and optimization**. Engineers run a variety of test programs and simulations to ensure that the machine is working correctly and can handle the massive computational loads it was designed for. These tests check for errors in the hardware, bottlenecks in the software, and inefficiencies in the cooling and power systems.

Benchmarking is a critical part of this process, as supercomputers are often ranked based on their performance in standard tests, such as the **LINPACK benchmark**, which measures how fast a machine can solve complex mathematical equations. Supercomputers are often compared globally, with rankings published on the **TOP500** list of the world's most powerful computers.

8. Applications and Real-World Impact

Once built, supercomputers serve a wide range of applications, from modeling climate change and predicting natural disasters to aiding in the development of new drugs and materials. They are indispensable tools for scientific research, helping to simulate physical phenomena like fluid dynamics, molecular interactions, and even the formation of galaxies.

Supercomputers also play a critical role in national defense, cryptography, and cybersecurity, where their immense processing power is used to break complex codes, simulate nuclear reactions, and safeguard digital infrastructure.

Conclusion

Building a supercomputer is a monumental task that requires a blend of cutting-edge technology, creative problem-solving, and collaboration across multiple disciplines. From the processors and nodes that form its core to the cooling systems that prevent overheating, every component of a supercomputer must be carefully engineered to ensure that it can operate at peak performance.

While supercomputers are primarily used for specialized scientific and industrial applications, their influence is felt throughout society, driving innovation in everything from medical research to artificial intelligence. As we continue to push the boundaries of what these machines can do, the future of supercomputing promises even greater breakthroughs in understanding the complex systems that govern our world.

How Are Video Games Developed?

Video game development is a complex, multi-faceted process that combines creativity, technology, and teamwork. It brings together diverse fields such as art, programming, storytelling, and sound design to create immersive, interactive worlds that entertain millions of players worldwide. But behind the scenes of every video game lies a highly structured process that turns initial concepts into polished, playable products. Let's break down the key stages of video game development and see how this magic happens.

1. Concept and Idea Development

Every game starts with an idea. This could come from a small indie developer or a large studio, but at the core is a unique concept—a game mechanic, story, or character that captures attention. During this stage, game designers brainstorm ideas, sketch initial characters, and map out the game's basic objectives.

The concept phase usually answers key questions:

- What is the core gameplay mechanic?
- Who is the target audience?

- What platform will the game be released on (PC, console, mobile)?

A **game design document** is often created to outline the gameplay, story, art style, and overall vision for the game.

2. Pre-Production

Once the concept is solidified, the pre-production phase begins. Here, the development team determines the technical aspects of the game, including the tools and software that will be used. Pre-production also involves prototyping, which allows developers to test game mechanics early in the process.

This phase often involves:

- **Prototyping the core mechanics** to see if they're fun or functional.
- **Selecting the game engine** (Unity, Unreal Engine, etc.) based on the needs of the project.
- **Creating a roadmap** for production timelines, budgets, and team roles.

A key part of pre-production is also storyboarding. For story-driven games, writers and designers collaborate to map out the game's plot and narrative structure.

3. Production

Behind the Scenes

This is the most extensive and labor-intensive phase, where the majority of the game is actually created. During production, multiple teams work together simultaneously on different aspects of the game:

- **Programming and Coding:** The backbone of any game, programmers build the actual code that makes the game function. They handle everything from physics, player controls, and AI behavior to creating the game engine if it's custom-built.
- **Game Design and Level Design:** Designers build the levels, ensuring that they are balanced, challenging, and aligned with the core mechanics. Level designers create the environments, pacing, and flow of the game experience.
- **Art and Animation:** Artists bring the visual world to life, designing characters, environments, objects, and special effects. Animators give life to characters, ensuring smooth movements, combat sequences, and interactions.
- **Sound and Music:** Audio engineers create sound effects, background music, and dialogue recordings that enhance immersion and gameplay. For large-scale games, soundtracks may be composed or licensed to match the game's tone.
- **Writing and Dialogue:** For narrative-heavy games, writers develop the script, dialogue, and branching storylines that add depth to the game. Games with

complex characters often require detailed scripts, sometimes even involving professional voice actors.
- **User Interface (UI) Design:** The UI team focuses on designing the menu systems, heads-up display (HUD), and control schemes, ensuring that the game is intuitive and accessible to players.

4. Testing

No game is complete without rigorous testing. Quality Assurance (QA) testers play through the game repeatedly, looking for bugs, glitches, and other issues. They also assess the game's balance, difficulty, and playability to ensure an optimal user experience. There are several types of testing:

- **Alpha Testing:** An early version of the game (usually with incomplete features) is tested internally to identify major issues.
- **Beta Testing:** A more polished version is shared with external players or the public to gather feedback, bug reports, and insights into user experience. Many developers release **closed** or **open beta versions** to get real-world feedback before the official release.
- **Stress Testing:** For multiplayer games, servers and online infrastructures are stress-tested to ensure they can handle large numbers of players.

Testing is an ongoing process, often continuing up until release day and beyond. The feedback from testers is crucial in polishing the final product.

5. Post-Production and Launch

Once the game is fully developed, polished, and tested, it moves into post-production. This phase focuses on marketing, final bug fixes, and preparing the game for release. Some key activities include:

- **Optimization:** Ensuring the game runs smoothly on all intended platforms, fixing bugs, and improving performance.
- **Marketing and Publicity:** Developers begin to promote the game through trailers, gameplay demos, interviews, and partnerships with gaming influencers and media outlets.
- **Final Build and Certification:** For console games, the final version must go through a certification process with platform holders like PlayStation, Xbox, or Nintendo to ensure it meets their technical standards.

Finally, the game is released to the public through various digital or physical distribution channels.

6. Post-Launch Support

Even after release, game development isn't truly finished. Most modern games require post-launch support, which may include:

- **Patches and Updates:** Developers release updates to fix bugs that were not caught before release, optimize performance, or address user feedback.
- **Downloadable Content (DLC):** Many developers create additional content, expansions, or new game modes to keep players engaged.
- **Live Services:** For online multiplayer games, servers are maintained, and regular updates or seasonal events are added to keep the community active.

Conclusion

Video game development is a long, intricate process that blends artistic vision, technical prowess, and player feedback to create immersive and engaging experiences. Behind every game you play lies countless hours of design, programming, and creativity. Whether you're battling monsters in an epic role-playing game, solving puzzles in a mobile game, or navigating vast open worlds, you're experiencing the result of meticulous planning and hard work from a dedicated team.

The next time you pick up a controller or tap on your smartphone screen, take a moment to appreciate the

Behind the Scenes

journey that game took from a simple idea to the finished product in your hands!

Behind the Scenes

How is Water Purified for Drinking?

Water is one of the most essential resources for human survival, yet not all water is safe to drink in its natural state. From rivers, lakes, and underground reservoirs to rainwater, the journey of transforming raw water into safe drinking water involves several complex purification processes. These methods are designed to remove contaminants, pathogens, and harmful substances to ensure the water is fit for human consumption. In this section, we'll explore the fascinating processes behind water purification, from traditional methods to advanced modern technologies.

The Sources of Water

Water used for drinking comes from various natural sources, including:

- **Surface water:** Rivers, lakes, and reservoirs
- **Groundwater:** Aquifers and wells
- **Rainwater:** Collected from rainfall
- **Desalinated water:** Ocean water that undergoes desalination (removal of salt)

Behind the Scenes

However, before this water can reach your tap, it must go through a purification process to remove impurities such as bacteria, viruses, dirt, chemicals, and pollutants.

The Step-by-Step Process of Water Purification

1. **Coagulation and Flocculation** The purification process begins with coagulation, where chemicals called coagulants (such as alum) are added to the water. These chemicals cause small particles like dirt, silt, and organic matter to clump together. This clumping process is known as **flocculation**, where the particles form larger clumps called "flocs." The flocs are heavy enough to settle out of the water, making it easier to remove.
2. **Sedimentation** Once flocculation is complete, the water enters a sedimentation tank where the flocs settle to the bottom. This step allows large particles to be separated from the cleaner water. The clearer water, now free of most large particles, moves on to the next stage of purification, while the settled material (sludge) is removed.
3. **Filtration** After sedimentation, the water passes through filters to remove remaining suspended particles, bacteria, and other microorganisms. These filters are usually made of layers of sand, gravel, and sometimes charcoal. The water flows through the filter bed, trapping fine particles in the process.

Behind the Scenes

Filtration not only removes contaminants but also improves the clarity and taste of the water.

4. **Disinfection** Although filtration removes most contaminants, harmful microorganisms like bacteria, viruses, and parasites may still remain in the water. To eliminate these pathogens, the water undergoes disinfection. The most common disinfection method is the addition of **chlorine** or **chloramine** (a combination of chlorine and ammonia), which effectively kills any remaining germs. In some areas, **ozone** or **ultraviolet (UV) light** is used to disinfect water without adding chemicals.

5. **pH Adjustment** To ensure the water is not too acidic or too basic, its **pH** is adjusted during purification. Adding substances like **lime** or **sodium hydroxide** neutralizes the acidity, preventing corrosion of pipes and ensuring the water is safe to drink. Proper pH levels also help maintain the effectiveness of chlorine as a disinfectant.

6. **Fluoridation** In many regions, fluoride is added to the water supply during the purification process. **Fluoridation** helps strengthen teeth and prevent dental cavities. The process is carefully controlled to ensure the fluoride level is safe and beneficial to human health.

7. **Distribution** Once the water has undergone all the purification steps, it is pumped into a network of

pipes that distribute it to homes, businesses, and other locations. Throughout the distribution process, the water is continuously monitored to ensure it remains free of contamination and meets quality standards.

Advanced Water Purification Techniques

In addition to the traditional methods described above, modern technologies have introduced advanced water purification techniques, especially in areas where freshwater is scarce or where water sources are highly contaminated:

- **Reverse Osmosis (RO):** This process involves pushing water through a semipermeable membrane to remove impurities such as salts, heavy metals, and chemicals. RO is widely used for desalinating seawater and purifying brackish water.
- **Ultraviolet (UV) Purification:** UV light is used to destroy harmful microorganisms in water. This chemical-free method is highly effective at killing bacteria and viruses, though it does not remove chemical contaminants.
- **Desalination:** For coastal areas or regions with access to seawater, desalination is a key process. **Thermal desalination** uses heat to evaporate water and separate it from salts, while **membrane**

desalination (like reverse osmosis) uses pressure to filter out salts. Both methods make seawater safe to drink.

- **Carbon Filtration:** Activated carbon filters are used to remove organic compounds, chlorine, and other chemicals that can affect the taste, odor, and safety of water. This method is commonly used in household water filtration systems.

Why Water Purification Is Vital

Water purification is essential to protect public health and ensure access to clean, safe drinking water. Without proper treatment, water can contain dangerous pathogens such as **E. coli, Giardia,** and **Cryptosporidium**, as well as harmful chemicals like **lead, arsenic**, and **pesticides**. These contaminants can lead to serious health problems, including gastrointestinal illnesses, reproductive issues, and neurological disorders.

In addition to health concerns, purified water is also necessary for various industrial and agricultural processes. Whether it's being used for irrigation, food production, or manufacturing, clean water is a critical component of many industries.

Conclusion: A Global Responsibility

Behind the Scenes

While many of us have access to clean drinking water, millions of people around the world still struggle with water scarcity and contamination. Water purification technologies continue to evolve as researchers develop more efficient and sustainable ways to provide clean water to those in need. From large-scale water treatment plants to portable filtration systems in disaster-stricken regions, the quest for clean water is an ongoing challenge for both governments and innovators.

Behind the Scenes

How is Cryptocurrency Mined?

Cryptocurrency mining is a process that powers the decentralized digital currency system, and it has become one of the most fascinating and complex aspects of the modern financial world. Mining serves as the backbone of many cryptocurrencies, such as Bitcoin, Ethereum, and others. But how exactly does this process work? What's happening behind the scenes when we talk about "mining" in a digital context? This chapter breaks down the steps, technology, and mathematical algorithms involved in cryptocurrency mining and explains how miners help secure the network and validate transactions.

The Basics of Cryptocurrency Mining

At its core, cryptocurrency mining is the process of verifying and adding transactions to the blockchain, the decentralized digital ledger that underpins cryptocurrencies. The blockchain records every transaction made with a cryptocurrency, ensuring that the same unit of digital currency is not spent more than once. Miners play a crucial role in this system by solving complex cryptographic puzzles to validate blocks of transactions, which are then added to the blockchain.

Behind the Scenes

In return for their work, miners are rewarded with newly created units of the cryptocurrency they are mining, along with transaction fees. This dual reward system incentivizes miners to continue securing the network and verifying transactions, which is crucial to the decentralized nature of cryptocurrencies.

How Mining Works: The Proof of Work Concept

The most common method of mining cryptocurrencies, particularly Bitcoin, is based on a consensus mechanism known as **Proof of Work (PoW)**. Proof of Work requires miners to compete with each other to solve a complex mathematical puzzle. This puzzle involves finding a specific hash—a string of numbers and letters—by running the block's data through a cryptographic hash function. The solution must meet a specific condition, such as starting with a certain number of zeros.

The cryptographic puzzle is designed to be difficult and resource-intensive to solve, but easy to verify. This ensures that miners have to expend computational power (work) to find the solution. Once a miner finds the correct hash, they broadcast it to the network, and other miners verify it. If the solution is correct, the miner who solved the puzzle earns a block reward, which typically includes new cryptocurrency tokens and transaction fees from the verified block.

Behind the Scenes

The Role of Mining Hardware

Cryptocurrency mining requires specialized hardware to perform the immense number of calculations needed to solve the cryptographic puzzles. Early on, Bitcoin miners used standard CPUs (Central Processing Units) from personal computers, but as the network grew, the difficulty of mining increased. This led to the adoption of **Graphics Processing Units (GPUs)**, which are much more efficient at handling the parallel calculations required for mining.

Today, the most competitive miners use **ASICs (Application-Specific Integrated Circuits)**—specialized machines designed solely for cryptocurrency mining. ASICs are incredibly powerful and energy-efficient compared to traditional hardware, but they are also expensive and consume vast amounts of electricity.

Mining Farms and Energy Consumption

As cryptocurrency mining became more competitive, many miners realized that running a single mining rig would not generate significant rewards. This led to the creation of **mining farms**—large-scale operations that house hundreds or thousands of mining rigs working in parallel. These mining farms can be found in countries where electricity costs are low, allowing miners to maximize profits.

However, the energy consumption of large-scale cryptocurrency mining has become a controversial topic. Bitcoin mining, for example, is notorious for its high energy requirements. The computational power required to solve Proof of Work puzzles consumes a massive amount of electricity, which has sparked concerns about the environmental impact of mining. Some cryptocurrencies, such as Ethereum, are transitioning to more energy-efficient consensus mechanisms like **Proof of Stake (PoS)** to address these concerns.

Difficulty Adjustments and Halving

Mining cryptocurrency isn't a static process. The difficulty of mining adjusts over time to ensure that new blocks are added to the blockchain at a regular interval. For Bitcoin, a new block is added roughly every 10 minutes. As more miners join the network and the total computational power increases, the difficulty of the cryptographic puzzle increases, making it harder to mine new blocks. Conversely, if miners leave the network, the difficulty decreases.

Another important aspect of mining is the **halving** event. For cryptocurrencies like Bitcoin, the block reward miners receive is halved approximately every four years. This process is built into Bitcoin's code to limit the total supply to 21 million bitcoins, making it a deflationary asset.

Halving reduces the rate at which new bitcoins are created, which can impact the profitability of mining and the overall value of the cryptocurrency.

Mining Pools: Strength in Numbers

For individual miners, the chances of solving a block on their own can be incredibly slim, especially as the network grows and difficulty increases. To improve their chances of earning rewards, many miners join **mining pools**—groups of miners who combine their computational power to increase their chances of solving the puzzle. When the pool successfully mines a block, the rewards are distributed among the participants based on the amount of work they contributed.

Mining pools are particularly important for smaller miners, as they allow them to participate in the network and earn consistent rewards, even if their individual computational power is limited.

Cloud Mining

Another option for individuals who want to get involved in cryptocurrency mining without purchasing expensive hardware is **cloud mining**. Cloud mining services allow users to rent computational power from remote data centers that perform the mining on their behalf. In exchange for a fee, users receive a portion of the rewards generated by the mining operation.

While cloud mining offers a more accessible way to participate in mining, it's important to carefully evaluate the legitimacy of cloud mining services. Some providers have been known to operate Ponzi schemes or fail to deliver promised returns.

Challenges and Risks of Cryptocurrency Mining

Cryptocurrency mining, while potentially profitable, comes with its share of challenges and risks:

1. **High Energy Costs:** Mining requires a significant amount of electricity, which can result in high operational costs. In some cases, the cost of electricity can outweigh the value of the cryptocurrency mined.
2. **Expensive Hardware:** ASICs and other mining equipment can be costly to purchase and maintain. Additionally, hardware can become obsolete as mining difficulty increases, requiring constant upgrades.
3. **Market Volatility:** Cryptocurrency prices are notoriously volatile. A sharp drop in the value of a cryptocurrency can drastically reduce the profitability of mining, making it a risky investment.

4. **Regulatory Uncertainty:** The legal and regulatory landscape surrounding cryptocurrency mining varies by country. In some regions, governments have imposed restrictions or outright bans on mining due to its environmental impact or association with illegal activities.

Conclusion

Cryptocurrency mining is a complex and evolving field that blends cutting-edge technology with economic incentives. It plays a crucial role in maintaining the decentralized nature of digital currencies and ensuring that transactions are secure and verified without the need for a central authority. As the world of cryptocurrency continues to grow and change, so too will the methods and technologies used in mining. While it remains a highly competitive and energy-intensive process, innovations such as Proof of Stake and renewable energy solutions offer a glimpse into a more sustainable future for cryptocurrency mining.

By understanding the inner workings of cryptocurrency mining, we gain a greater appreciation for the technology that powers the decentralized financial revolution. Whether you're a miner, investor, or simply a curious observer, the secrets behind cryptocurrency mining reveal

Behind the Scenes
a fascinating intersection of technology, economics, and innovation.

Behind the Scenes

How Are Ancient Artifacts Preserved?

Preserving ancient artifacts is a delicate and complex process that bridges the worlds of archaeology, science, and history. These priceless relics offer a window into past civilizations, but the ravages of time, exposure to the elements, and human interference can severely damage them. To ensure these objects remain intact for future generations, experts use a combination of advanced techniques, scientific methods, and painstaking care. Let's dive into the fascinating process of how ancient artifacts are preserved and protected.

1. Excavation and Initial Handling

Preservation starts the moment an artifact is unearthed during an archaeological excavation. The careful removal of artifacts from the ground is crucial, as improper handling can cause immediate damage. Archaeologists meticulously document the artifact's location, position, and surrounding conditions, as this information is essential for understanding the artifact's context.

Artifacts are then carefully cleaned, with great care taken to avoid causing any deterioration. This initial cleaning is usually done using gentle brushes or even air blowers to remove dirt and debris without harming the fragile surface of the object. In some cases, if the artifact is too delicate, it may be encased in a protective mold and transported to a conservation lab before further cleaning.

2. Preventing Deterioration

One of the greatest threats to ancient artifacts is the environment. Exposure to air, moisture, and temperature fluctuations can accelerate the degradation process. To combat this, conservators often stabilize the artifacts in controlled environments.

For example:

- **Temperature and Humidity Control:** In museums or storage facilities, temperature and humidity are carefully regulated. Most artifacts need a stable, cool, and slightly humid environment to prevent cracking, warping, or disintegration. Materials like wood, textiles, and paper are especially sensitive to changes in humidity.
- **Protection from Light:** Ultraviolet (UV) light from the sun or artificial lighting can cause colors to fade and materials like cloth, paint, and even stone to degrade. Artifacts are often displayed in low-light environments or with UV-filtering glass to protect them from harmful radiation.

Behind the Scenes

3. Chemical Conservation Techniques

Many ancient artifacts, particularly those made from metals, can suffer from corrosion or other chemical reactions over time. For example, iron rusts when exposed to moisture, and copper-based artifacts, such as bronze, can develop a green patina. While some surface changes are natural and even desirable (like the patina on a bronze sculpture), excessive corrosion can weaken the structure of the artifact.

To prevent further degradation:

- **Chemical Treatments** are applied to stabilize the material. Conservators may use chemical baths to remove corrosion or apply protective coatings to prevent future damage. For example, iron artifacts might be soaked in a solution that removes rust, while bronze objects might be treated with wax or lacquer to shield them from the air.
- **Desalination** is another important process, especially for artifacts recovered from underwater. When items are submerged for long periods, salt deposits can form inside the material. If not removed, these salts can crystallize and cause cracks or flaking. Desalination involves soaking the object in fresh water to slowly extract the salts over time.

4. Repair and Reconstruction

In some cases, artifacts are discovered in pieces or have been damaged over the years. Conservators use a variety of techniques to repair and reconstruct these objects while maintaining their authenticity. However, the goal is never to make an artifact look "brand new" but to preserve its original form as much as possible.

- **3D Scanning and Modeling:** Modern technology allows for the creation of digital models of artifacts, which can then be used to recreate missing pieces. For instance, if a fragment of a vase is missing, 3D scanning can help model the missing portion, which can then be carefully recreated and attached to the original.
- **Traditional Materials:** When repairing ancient artifacts, conservators try to use materials that are similar to the original, such as natural resins or adhesives. However, they are careful to ensure that any repairs can be easily identified and undone if future technology offers better solutions.

5. Long-Term Storage and Monitoring

Preserving ancient artifacts doesn't end with stabilization and repair. Long-term preservation requires continuous monitoring and maintenance to ensure that environmental conditions remain optimal and that the artifact isn't deteriorating over time. Many museums and storage

Behind the Scenes

facilities employ sophisticated monitoring systems to track changes in temperature, humidity, and even air quality.

- **Storage Materials:** Artifacts are stored in specialized containers, boxes, or shelves designed to minimize contact with harmful materials. Acid-free paper, padded boxes, and inert plastics are commonly used to prevent chemical reactions that could damage the artifact.
- **Regular Condition Assessments:** Conservators routinely inspect artifacts for any signs of new damage. If an artifact shows signs of deterioration, it can be removed from display and treated before the damage becomes irreversible.

6. Documentation and Digital Preservation

Every artifact is meticulously documented throughout its conservation process. High-resolution photography, detailed written reports, and digital scans create a permanent record of the object's condition and any treatments it undergoes. This documentation is invaluable for future conservators, researchers, and historians.

In recent years, **digital preservation** has become an increasingly important tool in artifact conservation. Technologies such as 3D scanning and virtual reality are used to create digital replicas of artifacts, allowing people to study them in detail without physically handling the original objects. Digital preservation ensures that even if an artifact is lost or damaged in the future, a detailed record of its existence remains.

7. Ethical Considerations

Preserving ancient artifacts comes with ethical responsibilities. Conservators must balance the need to protect an object with the obligation to preserve its historical integrity. This often means that treatments must be as non-invasive as possible, and any modifications or repairs should be reversible. Additionally, cultural sensitivity is paramount, especially when dealing with sacred or culturally significant objects. Conservators work closely with archaeologists, historians, and sometimes representatives from the originating culture to ensure that preservation efforts respect the artifact's heritage.

Conclusion

Preserving ancient artifacts is a blend of science, art, and ethics. It requires not only technical expertise but also a deep respect for the history and culture embodied in these objects. Every artifact tells a story, and through careful conservation, we ensure that these stories endure for generations to come.

As technology advances and our understanding of materials improves, so too do our methods of preservation.

Behind the Scenes

From the moment an artifact is discovered to the years it spends on display or in storage, the work of conservators is never done. Their efforts allow us to maintain a tangible connection to the past, safeguarding the world's cultural heritage one artifact at a time.

Behind the Scenes

How is Waste Managed in Space?

Space exploration presents extraordinary challenges, and one often-overlooked issue is waste management. On Earth, waste disposal is routine, with well-established systems for handling everything from household garbage to industrial pollutants. In space, however, managing waste is a far more complex task due to the unique conditions of microgravity, limited space, and the need for sustainable systems on long-duration missions.

The Problem of Space Waste

Astronauts living aboard the International Space Station (ISS) generate waste just like they would on Earth, including food packaging, human waste, and various materials used during scientific experiments. But in the confined environment of a spacecraft, waste accumulation poses health risks and can interfere with equipment if not properly managed.

Additionally, beyond the personal waste generated by astronauts, space missions themselves contribute to **space debris**—defunct satellites, rocket stages, and small fragments that pose hazards to active spacecraft and satellites. This form of "space junk" is a growing concern as more missions are launched, creating risks of collisions in orbit.

Waste Disposal Systems in Space

Managing waste on the ISS and future spacecraft involves several strategies that must account for resource efficiency, weight constraints, and safety.

1. **Collection and Compression:** Waste generated aboard spacecraft is collected and compressed to reduce volume. Special containers are used to store trash in sealed compartments. These containers are designed to ensure that waste does not release harmful gases or contaminate the space environment.
2. **Human Waste Disposal:** Managing human waste in space is one of the more challenging tasks. The ISS employs advanced toilet systems that vacuum liquid and solid waste into separate containers. Liquid waste, primarily urine, is treated and **recycled** into drinking water through sophisticated filtration systems, which is crucial for water conservation during long missions.

   Solid waste, however, cannot yet be recycled in space. Instead, it is stored in containers and placed on cargo ships, like the Cygnus spacecraft, which periodically detach from the ISS and burn up in

Behind the Scenes

Earth's atmosphere upon re-entry. This effectively disposes of the waste without leaving traces in orbit.

3. **Recycling and Resource Recovery:** On longer missions—such as potential trips to Mars—space agencies are developing systems that go beyond waste disposal, aiming for waste **recycling and resource recovery**. Future spacecraft may use technologies to convert solid waste into useful materials. For example, NASA has been testing systems that can turn human waste into methane, which can be used as rocket fuel or to generate energy.

   Advanced bioreactors are also being explored to break down waste biologically and produce oxygen or water. This will be critical for supporting long-term life support systems where resupply missions from Earth are not feasible.

Space Debris: The Bigger Issue

While waste management inside spacecraft is essential for astronaut health, a more pressing problem is the growing accumulation of **space debris**. Thousands of pieces of orbital debris—including abandoned satellites and fragments from disintegrated spacecraft—are circling the Earth at high velocities. Even small objects traveling at such high speeds can cause catastrophic damage to active spacecraft or satellites.

1. **Tracking and Avoidance:** Space agencies like NASA and ESA (European Space Agency) actively track space debris using radar and telescopic systems to predict potential collisions. When necessary, the ISS and other spacecraft are equipped to perform **evasive maneuvers** to avoid dangerous debris.
2. **Mitigation and Removal:** To reduce the growth of space debris, new regulations require satellites and rockets to be designed for **end-of-life disposal**, meaning they must either re-enter Earth's atmosphere and burn up or be moved into a "graveyard orbit" where they pose less of a collision risk.

   Additionally, various methods for actively removing space debris are being researched. These include using harpoons, nets, and robotic arms to capture and deorbit larger pieces of debris, as well as utilizing lasers to push smaller fragments out of harm's way.

Future Innovations in Space Waste Management

As humanity sets its sights on deep space exploration, including missions to Mars, the Moon, and beyond, waste management systems will need to become even more sophisticated. Long-duration missions will require **closed-loop life support systems**, where waste is treated as a resource rather than discarded. This means developing technologies that can fully recycle human waste into water, oxygen, or fuel.

NASA is already working on initiatives such as the **"Trash-to-Gas"** concept, which could convert solid waste into methane fuel. Similarly, the European Space Agency is exploring the use of microbes in bioreactors to process organic waste into usable products.

In the far future, advanced nanotechnology or even **artificial photosynthesis** could help astronauts sustainably manage waste while producing vital resources like food and oxygen.

Conclusion

Waste management in space is a highly complex challenge that requires innovative solutions. From dealing with human waste aboard the ISS to tackling the growing issue of space debris, space agencies are continuously advancing their strategies to ensure that astronauts can live and work safely in space. As space exploration continues to push new boundaries, the need for sustainable, efficient waste management systems will only grow, driving further technological advancements that could one day benefit life on Earth as well.

Behind the Scenes

How Is Artificial Meat Grown in Labs?

Artificial meat, also known as **lab-grown meat** or **cultured meat**, represents one of the most innovative breakthroughs in food technology. Unlike traditional meat, which comes from slaughtering animals, lab-grown meat is produced by cultivating animal cells in a controlled environment. This process not only offers a more sustainable and ethical alternative to conventional meat production but also holds the potential to revolutionize the global food industry. But how exactly is artificial meat grown in labs? Let's take a look behind the scenes at the fascinating science and technology involved in creating this futuristic food.

The Science of Cultured Meat

At its core, lab-grown meat is based on **cellular agriculture**—a method that uses animal cells as the starting material and grows them in a laboratory setting, rather than raising and slaughtering livestock. Here's how it's done step by step:

1. **Cell Sourcing:** The first step in the process of creating artificial meat involves obtaining a small sample of animal cells. These can be taken from a

live animal through a biopsy, usually without harming the animal. The cells that are selected are typically **muscle cells**, but **fat cells** and **connective tissue cells** can also be used to give the final product a more authentic texture and taste.
2. **Stem Cell Selection:** Among the cells sourced, scientists often focus on **stem cells**. Stem cells are special because they can differentiate into various types of cells, including muscle cells, which are the main component of meat. These cells are isolated and encouraged to multiply in the lab, creating the building blocks for the meat-growing process.
3. **Cell Cultivation:** Once the stem cells are harvested, they are placed into a **bioreactor**—a special device that provides the ideal conditions for cells to grow. The bioreactor is filled with a **growth medium**, which acts as a nutrient-rich liquid that feeds the cells. The growth medium contains essential nutrients such as amino acids, sugars, vitamins, and minerals, similar to what cells would receive in a living animal's body.

This stage is crucial because the cells need a precise balance of nutrients and environmental conditions (like temperature and pH) to grow and multiply effectively. The cells divide and multiply rapidly, forming larger masses of muscle tissue over time.

4. **Scaffolding for Structure:** One challenge in producing lab-grown meat is mimicking the texture and structure of real meat. To address this, scientists use **scaffolds**—biodegradable materials that provide a framework for the cells to grow on. Scaffolds help guide the cells to form the desired shapes, textures, and muscle structures. For example, these scaffolds can simulate the alignment of muscle fibers found in a steak, giving the lab-grown meat a similar bite and feel to traditionally farmed meat.
5. **Tissue Formation:** As the cells continue to multiply and grow on the scaffolds, they eventually form muscle tissues. These tissues are the primary component of meat. Over time, the cells mature, and with the right combination of nutrients and growth factors, they begin to develop into larger, structured pieces of meat.
6. **Fat and Flavor Development:** To enhance the flavor and mouthfeel of lab-grown meat, fat cells are often added during the growing process. Fat contributes to the richness and juiciness of meat, and scientists can control the ratio of muscle to fat to replicate different cuts of meat. Additionally, other natural additives like **flavor compounds** and **coloring agents** may be included to further enhance the appearance and taste.

Behind the Scenes

7. **Harvesting the Meat:** Once the muscle tissues have reached the desired size and maturity, they are harvested from the bioreactor. At this stage, the meat is processed similarly to traditional meat, where it can be formed into various products such as burgers, steaks, or sausages. Because lab-grown meat is cultivated in a controlled environment, it can be produced with a lower risk of contamination from bacteria like E. coli or Salmonella.
8. **Final Product:** The harvested meat is ready to be cooked and consumed. Since it is real animal meat at the cellular level, lab-grown meat is virtually indistinguishable from conventional meat in terms of its biological makeup. However, it can be customized to improve nutritional content, texture, or even enhance certain flavors, offering endless possibilities for innovation in the food industry.

Benefits of Lab-Grown Meat

The production of artificial meat offers several compelling benefits:

- **Sustainability:** Traditional meat production is resource-intensive, requiring vast amounts of land, water, and feed, while contributing to greenhouse gas emissions. Lab-grown meat, on the other hand,

requires significantly fewer resources and produces a fraction of the environmental impact.
- **Animal Welfare:** Cultured meat eliminates the need for large-scale animal farming and slaughter, addressing ethical concerns related to animal welfare. Animals do not need to be killed to produce lab-grown meat, and the process can be done with minimal harm to the donor animals.
- **Food Security:** As the global population continues to grow, the demand for meat is expected to rise. Lab-grown meat offers a solution to food shortages by providing a scalable and reliable alternative to traditional livestock farming, potentially addressing global hunger.
- **Health and Safety:** Because lab-grown meat is produced in sterile, controlled environments, it can be engineered to be free from antibiotics, hormones, and pathogens, making it a healthier and safer option for consumers.

Challenges and Future of Lab-Grown Meat

While lab-grown meat presents exciting possibilities, there are still challenges to overcome before it becomes a mainstream food source:

- **Cost:** The production of lab-grown meat is still more expensive than conventional meat, although

advancements in technology are steadily driving costs down. Researchers are working on optimizing the growth medium and bioreactors to make the process more cost-efficient.
- **Consumer Acceptance:** While many people are intrigued by the idea of lab-grown meat, widespread consumer acceptance is still in the early stages. Public perception, ethical considerations, and taste preferences will play a critical role in determining the future success of this technology.
- **Regulation and Scaling:** As with any new food technology, lab-grown meat must undergo rigorous testing and regulation before it can be widely distributed. Scaling up production to meet global demand will also require significant infrastructure and investment.

Conclusion

Artificial meat is no longer a futuristic concept—it's becoming a reality. Through innovative science, careful cultivation, and meticulous attention to detail, lab-grown meat is paving the way for a more sustainable, ethical, and efficient food system. While there are still hurdles to overcome, the potential benefits are enormous. Whether we see lab-grown steaks on supermarket shelves in the next few years or gourmet cultured burgers in high-end restaurants, one thing is clear: lab-grown meat could

Behind the Scenes

change the way we think about food and how we consume it.

By understanding the process of how lab-grown meat is made, we gain insights into the future of food production—one that could reshape our diets, our environment, and our relationship with the natural world.

Behind the Scenes

How Are Films Animated?

Animation is one of the most captivating and versatile forms of storytelling, capable of bringing fantastical worlds, characters, and ideas to life in ways that live-action films cannot. Whether it's the charming hand-drawn classics of early Disney films, the complex stop-motion of movies like *Coraline*, or the dazzling computer-generated imagery (CGI) seen in modern blockbusters, the process of creating animated films is a blend of artistry, technology, and meticulous attention to detail. But how exactly are these incredible films made? Let's dive into the process of film animation, exploring the different techniques and the intricate steps involved in crafting animated movies.

1. Concept and Storyboarding

The animation process begins like any other film: with a story. Writers, directors, and creative teams come together to develop the narrative, characters, and themes. Once the story is outlined, the next step is **storyboarding**.

Storyboards are like a visual script for the movie, where each scene is sketched out in sequence. These sketches show how the characters and objects will move, interact, and appear within each shot. This step is crucial because it acts as a blueprint for the entire film, helping animators, directors, and producers visualize how the final product will look. Storyboarding also allows for changes to be made early in the process before more time-consuming and expensive animation work begins.

2. Character Design and Model Sheets

Once the story and storyboard are in place, **character designers** take on the task of creating the visual identity of the characters. In traditional 2D animation, this involves drawing different views of each character to ensure consistency throughout the film. In CGI animation, 3D models are created, detailing the character's appearance, texture, and how they will move within a three-dimensional space.

Model sheets are created, which show the character from different angles and with various facial expressions. These sheets act as a reference for animators to keep the character's appearance uniform throughout the animation process. Every detail, from how a character's hair moves to how they smile, must be carefully designed.

3. Animatic Creation

The **animatic** is a rough version of the movie created using the storyboards. This version includes simple movement, timing, and rough audio. It allows the filmmakers to get a better sense of the pacing and how the scenes will flow together.

The animatic helps refine the timing of key scenes, comedic moments, action sequences, and transitions. It acts as a test run for the movie before diving into full animation. Changes are often made during this phase to ensure the film's rhythm feels right.

4. Animation Techniques

There are several different techniques used in animation, each with its own style, tools, and challenges:

2D Animation (Traditional Animation)

In traditional hand-drawn animation, every frame of movement is drawn by hand. Early films like *Snow White and the Seven Dwarfs* (1937) and *The Lion King* (1994) used this method, which required animators to create thousands of individual drawings. Each second of film typically requires 24 frames, meaning an animator would need to create 24 different drawings for every second of the movie.

Today, traditional animation has been largely replaced by digital tools, but the essence remains the same. Digital drawing tablets allow artists to create hand-drawn animation more efficiently, without needing physical paper.

Stop-Motion Animation

Behind the Scenes

In **stop-motion animation**, physical models or puppets are moved in small increments between individual photographs. When these photos are played in sequence, the objects appear to move on their own. Films like *The Nightmare Before Christmas* (1993) and *Kubo and the Two Strings* (2016) use this technique, which is painstakingly slow but offers a unique, tactile feel that CGI cannot replicate.

Creating stop-motion animation requires patience and precision. Every object, set, and character is handcrafted, and the smallest movement must be meticulously planned and captured.

3D CGI Animation

The most common form of animation today, **3D computer-generated animation**, is used in films like *Toy Story* (1995), *Frozen* (2013), and *The Incredibles* (2004). In CGI, characters and environments are modeled in three dimensions using software like Maya or Blender.

Once the models are created, animators can manipulate these 3D figures, adding motion and expression. One of the key advantages of CGI is the ability to create realistic textures, lighting effects, and environments, giving films a polished, life-like quality.

Behind the Scenes

3D animation often requires a large team of specialists, including modelers, riggers, animators, and texture artists, all working together to bring the characters and world to life.

5. Rigging and Motion

In both 2D and 3D animation, **rigging** is a vital step. A **rig** is like the skeleton of a character that determines how it moves. In 3D animation, riggers create a system of bones and joints that animators can manipulate to move the character.

Rigging defines how a character bends, stretches, and interacts with its environment. It also plays a key role in facial animations, allowing characters to display a wide range of emotions. Without proper rigging, characters would move in awkward, unrealistic ways.

6. Animating the Scenes

Once the rigs are in place, animators begin the labor-intensive process of **animating the scenes**. This involves moving the character frame by frame to create fluid, lifelike movement. Animators must have a deep understanding of anatomy, weight, timing, and expression to make their characters feel alive.

For instance, when a character walks across a room, animators must decide how their weight shifts, how fast they move, and what expressions are on their face. In action scenes, the animation must capture the energy, speed, and impact of movements.

In some cases, animators use **motion capture** to record live actors' movements, which are then translated into the animated characters. This technique was famously used in films like *Avatar* (2009) and *The Lord of the Rings* (2001).

7. Lighting, Texturing, and Rendering

After the animation is complete, the scenes need to be textured and lit. In CGI animation, **texture artists** add detail to surfaces, making them look smooth, rough, shiny, or matte, depending on the material.

Lighting artists simulate how light would behave in each scene, setting the mood and tone. Whether it's the warm glow of a sunset or the eerie light of a moonlit forest, lighting is key to creating atmosphere.

Finally, the film goes through the **rendering** process. Rendering is the final step where the animation is compiled and turned into its finished form. It can be a slow process—sometimes rendering a single frame can take several hours depending on the complexity of the scene.

8. Sound Design and Voice Acting

While the visuals are being developed, sound designers work on creating the audio for the film. **Voice actors** record the dialogue, often giving animators cues to match the characters' lip movements.

Sound effects, ambient noise, and a musical score are added to enhance the emotional depth of the film. The sound design plays a crucial role in immersing the audience in the animated world.

9. Post-Production and Final Touches

Once all the animation and sound work are complete, the film enters **post-production**. This phase includes adding final effects, such as special visual effects (VFX), color grading, and fine-tuning the audio mix. The film is carefully reviewed, and any last-minute changes are made before the final version is completed.

Conclusion

Creating an animated film is a highly collaborative and multi-layered process that combines artistic vision with technological expertise. Whether it's a hand-drawn 2D short or a big-budget CGI blockbuster, animation involves countless hours of work from a team of passionate artists and technicians.

Behind the Scenes

From concept to completion, each stage of animation is critical to bringing stories to life on the big screen. The next time you watch an animated film, you'll have a deeper appreciation for the incredible craftsmanship, patience, and creativity that go into making these magical worlds a reality.

Behind the Scenes

How is Nuclear Energy Produced?

Nuclear energy is one of the most powerful and efficient forms of energy production known to humanity. It harnesses the immense energy locked within the nucleus of atoms, specifically through a process called **nuclear fission**. By splitting heavy atomic nuclei, nuclear reactors release vast amounts of energy in the form of heat, which is then converted into electricity. Although nuclear energy has been a source of debate due to safety and environmental concerns, it remains a significant and clean alternative to fossil fuels.

The Process of Nuclear Fission

Nuclear energy is primarily produced through **nuclear fission**, a process in which the nucleus of a heavy atom, such as uranium-235 or plutonium-239, is split into smaller parts. This splitting releases an enormous amount of energy.

1. **Fuel Preparation**: The fuel used in nuclear reactors is typically uranium, which is mined and processed into pellets. These pellets are assembled into fuel rods, which are then grouped into fuel assemblies and placed inside the reactor core.

Behind the Scenes

2. **Neutron Bombardment**: Once the reactor is operational, neutrons are fired at the uranium or plutonium atoms inside the fuel rods. When a neutron hits the nucleus of a uranium-235 atom, the nucleus becomes unstable and splits into two smaller nuclei, releasing energy in the form of heat. This process also releases more neutrons, which go on to split more uranium nuclei in a self-sustaining chain reaction.
3. **Heat Generation**: As the uranium atoms split and the chain reaction continues, an immense amount of heat is generated. This heat is absorbed by a coolant, usually water, which is circulated through the reactor core to prevent overheating.
4. **Heat to Electricity Conversion**: The heated coolant, now turned into steam, is used to turn turbines connected to generators. As the turbines spin, they generate electricity. This is similar to how conventional power plants operate, but instead of burning coal or natural gas, the heat source is the nuclear fission reaction.

The Role of the Reactor Core

At the heart of a nuclear power plant is the **reactor core**, where the nuclear fission process takes place. The core contains the fuel assemblies, control rods, and coolant system. Control rods, made of materials like boron or cadmium, are inserted into the reactor to absorb excess neutrons and regulate the fission reaction. By adjusting the position of the control rods, operators can control the rate of the nuclear reaction, ensuring it remains steady and safe.

Types of Nuclear Reactors

There are different types of nuclear reactors, but the most common are:

1. **Pressurized Water Reactors (PWRs)**: In a PWR, water is heated by the nuclear reaction but kept under pressure so it doesn't boil. This superheated water is then transferred to a secondary loop where it turns to steam, which drives the turbines.
2. **Boiling Water Reactors (BWRs)**: In this design, water is allowed to boil directly inside the reactor core, and the steam produced drives the turbines.
3. **Fast Breeder Reactors (FBRs)**: These reactors generate more fissile material than they consume by converting uranium-238 into plutonium-239, making them highly efficient.

Safety Mechanisms and Concerns

Safety is a critical concern in nuclear energy production due to the potential for catastrophic accidents, as seen in

Behind the Scenes

Chernobyl and Fukushima. Modern reactors are designed with multiple **safety systems** to prevent such incidents:

- **Containment Structures**: Reactors are housed in massive containment buildings made of thick concrete and steel to prevent the release of radioactive material in case of an accident.
- **Emergency Cooling Systems**: In case of a failure in the primary cooling system, emergency cooling systems can rapidly cool down the reactor core to prevent overheating or a meltdown.
- **Automatic Shutdown**: If an issue is detected, modern reactors can automatically shut down by fully inserting the control rods to stop the fission reaction immediately.

Nuclear Energy: Benefits and Challenges

Advantages:

- **Low Carbon Emissions**: Unlike fossil fuel-based power plants, nuclear reactors produce very little carbon dioxide, making them a cleaner option for electricity generation.
- **High Energy Density**: A small amount of nuclear fuel can produce a vast amount of energy. For example, one uranium pellet (about the size of a fingertip) can produce as much energy as a ton of coal.

Challenges:

- **Radioactive Waste**: The spent fuel from nuclear reactors remains highly radioactive and dangerous for thousands of years, requiring secure storage solutions like deep geological repositories.
- **Nuclear Accidents**: Although rare, accidents at nuclear power plants can have devastating consequences, as seen in Chernobyl (1986) and Fukushima (2011).
- **Nuclear Proliferation**: The technology used for energy production can also be used to create nuclear weapons, raising concerns about proliferation and global security.

The Future of Nuclear Energy

Despite the challenges, nuclear energy is seen as a key component of the transition to a low-carbon future. Innovations such as **small modular reactors (SMRs)**, which are more flexible and easier to deploy, and **fusion energy**, which promises unlimited clean energy, are areas of active research and development. Additionally, there is a growing interest in improving nuclear waste disposal techniques and developing reactors that can use spent fuel as a new energy source.

Nuclear power has the potential to play a significant role in meeting the world's growing energy demands while combating climate change. However, it requires careful management, robust safety protocols, and ongoing technological innovation to ensure it remains a safe and sustainable option.

In summary, nuclear energy is produced through a complex process of splitting atomic nuclei, releasing vast amounts of energy. While the technology offers numerous advantages, it also comes with significant challenges, especially regarding safety and waste management. Understanding how nuclear energy is made and managed is crucial for grasping its potential role in the future of global energy.

How Is Virtual Money Regulated?

In recent years, the rise of virtual money—commonly known as cryptocurrency—has transformed the landscape of finance and commerce. Cryptocurrencies, such as Bitcoin, Ethereum, and countless others, have gained popularity for their potential to operate outside traditional banking systems, offering users a decentralized and often anonymous means of conducting transactions. However, as these digital currencies become more mainstream, the question of regulation arises. How is virtual money regulated, and what challenges do regulators face in ensuring the safety and integrity of this new financial frontier?

Understanding Virtual Money

Virtual money refers to digital currencies that exist only in electronic form and utilize cryptography for security. Unlike traditional currencies issued by governments (fiat money), virtual currencies operate on decentralized networks called blockchains. These networks enable secure peer-to-peer transactions without the need for intermediaries, such as banks. While this decentralization appeals to many users, it also raises concerns about fraud, money laundering, and market volatility.

The Need for Regulation

Behind the Scenes

As virtual money becomes increasingly integrated into the global economy, there are several compelling reasons for regulation:

1. **Consumer Protection**: With the growing popularity of cryptocurrencies, consumers are at risk of scams, fraud, and losses due to market volatility. Regulation can help establish safeguards to protect users from these risks.
2. **Preventing Financial Crime**: Cryptocurrencies can be used for illicit activities, including money laundering, tax evasion, and funding terrorism. Regulatory frameworks can help law enforcement agencies track and prevent such activities.
3. **Market Stability**: The cryptocurrency market is known for its volatility, which can lead to significant financial losses for investors. Regulation can introduce measures to promote market stability and reduce speculative trading.
4. **Tax Compliance**: As governments seek to collect taxes on digital asset transactions, regulation can provide clarity on tax obligations and reporting requirements for individuals and businesses.

Current Regulatory Approaches

Regulation of virtual money varies significantly across countries, with some embracing cryptocurrencies and others imposing strict restrictions. Here are some of the key regulatory approaches currently in place:

1. **Securities Regulation**: In many jurisdictions, cryptocurrencies that function as investment contracts are classified as securities and are subject to securities laws. For example, the U.S. Securities and Exchange Commission (SEC) has taken action against initial coin offerings (ICOs) that failed to comply with securities regulations.
2. **Anti-Money Laundering (AML) and Know Your Customer (KYC)**: Many countries require cryptocurrency exchanges and wallet providers to implement AML and KYC regulations. This means they must verify the identities of their users and report suspicious transactions to authorities, similar to traditional financial institutions.
3. **Tax Regulation**: Countries like the United States have established guidelines for taxing cryptocurrency transactions. The Internal Revenue Service (IRS) considers virtual currencies as property for tax purposes, requiring individuals to report capital gains or losses when they sell or exchange cryptocurrencies.
4. **Licensing and Registration**: Some countries have implemented licensing requirements for cryptocurrency exchanges and service providers. For instance, in Japan, exchanges must register with

the Financial Services Agency (FSA) and comply with strict operational standards.
5. **Central Bank Digital Currencies (CBDCs)**: In response to the rise of cryptocurrencies, some governments are exploring the development of their own digital currencies, known as central bank digital currencies (CBDCs). These state-backed currencies aim to combine the benefits of digital money with the stability of traditional fiat currency, while still allowing for regulatory oversight.

Challenges of Regulation

Despite the need for regulation, there are several challenges that regulators face:

1. **Rapid Technological Change**: The cryptocurrency landscape is continually evolving, with new technologies and concepts emerging regularly. This rapid pace makes it challenging for regulators to keep up and implement effective regulations.
2. **Global Nature of Cryptocurrencies**: Cryptocurrencies operate on a global scale, transcending national borders. This presents challenges for regulators, as inconsistent regulations across countries can create loopholes and opportunities for exploitation.

3. **Decentralization and Anonymity**: The very nature of cryptocurrencies—decentralized and often anonymous—makes it difficult to enforce regulations effectively. Regulators may struggle to track transactions and identify users involved in illicit activities.
4. **Innovation vs. Regulation**: Striking the right balance between fostering innovation and protecting consumers is a significant challenge. Overly strict regulations could stifle technological advancement and drive users to unregulated markets.

The Future of Virtual Money Regulation

As the cryptocurrency market matures, regulatory frameworks will likely continue to evolve. Collaboration between governments, regulatory bodies, and industry stakeholders will be essential in developing comprehensive regulations that promote safety, security, and innovation.

The future of virtual money regulation may involve:

- **International Cooperation**: Global coordination among regulators to establish common standards and practices for cryptocurrency regulation.
- **Adaptive Regulatory Frameworks**: Creating flexible regulations that can adapt to technological

advancements while ensuring consumer protection and financial integrity.
- **Enhanced Transparency**: Encouraging transparency in cryptocurrency transactions to foster trust among users and regulatory bodies.

Conclusion

Virtual money has the potential to revolutionize the financial landscape, but its rapid growth and unique characteristics present significant regulatory challenges. By implementing effective regulatory frameworks, governments can help ensure the safety, security, and integrity of virtual currencies, paving the way for a more stable and trustworthy financial future. As we navigate this exciting new frontier, it is essential to recognize that the regulation of virtual money is not just a legal obligation but a necessary step toward a sustainable digital economy that benefits all users.

Conclusion

As we reach the end of **"Behind the Scenes: Secrets of How Things Are Made – Book One,"** we have explored a vast array of industries, processes, and innovations that define the modern world. From the engineering marvels that shape our cities to the cutting-edge technologies that power our daily lives, this book has uncovered the incredible stories of how things are made, revealing the ingenuity, dedication, and collaboration behind these creations.

The Hidden Complexity Behind Everyday Life

One of the key takeaways from this book is the sheer complexity behind everyday objects and systems that we often take for granted. Whether it's the food we eat, the clothes we wear, or the smartphones we use, each product involves a series of carefully orchestrated steps, often involving a global network of designers, engineers, and manufacturers.

This complexity reflects the incredible advancements humans have made over centuries in perfecting methods of production, improving efficiency, and integrating new technologies. As we move forward, this evolution continues, driven by the desire to meet new challenges and improve upon what already exists.

For example, we've seen how the construction of a skyscraper requires not only brilliant architectural design but also precise engineering, innovative materials, and rigorous safety standards. Similarly, we've learned that producing something as seemingly simple as chocolate involves a sophisticated understanding of agriculture, chemistry, and industrial processes.

The Power of Innovation and Human Ingenuity

Innovation is a recurring theme throughout this book, and it is the driving force behind every advancement we've discussed. It's clear that nothing in our world stands still; industries are constantly evolving, adapting to new technologies and improving processes to meet the ever-changing demands of society.

The creation of electric vehicles, for instance, is a testament to how industries can pivot and innovate in response to global challenges like climate change. The evolution of the internet and digital communication technologies has completely transformed the way we connect, work, and even entertain ourselves. In all of these cases, it is human ingenuity—the relentless pursuit of new solutions—that has made progress possible.

This spirit of innovation is not confined to high-tech industries or large-scale engineering projects. It can be found in the craftsmanship of artisans, the creativity of designers, and the meticulous research of scientists. From 3D printing in manufacturing to AI-powered algorithms in healthcare, the potential for innovation knows no bounds, and the stories in this book illustrate just how far we've come—and where we might be headed next.

The Interconnected Nature of Our World

Another essential insight we've uncovered is the interconnectedness of various fields and industries. No product, system, or service exists in isolation. The construction of an airplane, for instance, requires the collaboration of engineers, chemists, materials scientists, safety experts, and even data analysts. Similarly, the production of medicine relies on biologists, chemists, regulatory bodies, and logistics companies to bring a life-saving drug from the lab to the patient.

As global supply chains become more complex, industries and technologies are increasingly interdependent. This interconnectedness highlights the importance of collaboration and communication across sectors to ensure that processes run smoothly and that innovations continue to advance. It's a reminder that while individual breakthroughs are exciting, it's the combined efforts of countless individuals and disciplines that lead to the creation of truly remarkable products.

Behind the Scenes

Looking to the Future: Sustainability and the Role of Innovation

As we reflect on the processes behind how things are made, it is crucial to consider how these industries and technologies are evolving to address the pressing challenges of our time—chief among them, sustainability. In an era of increasing environmental concerns, industries are being forced to rethink traditional production methods and move toward more sustainable practices.

From renewable energy to sustainable agriculture, the future of manufacturing and production lies in the ability to innovate in ways that are not only efficient but also environmentally responsible. Industries across the globe are developing new materials, reducing waste, and finding alternative energy sources to power their operations. For instance, the development of biodegradable packaging materials, energy-efficient manufacturing processes, and the increased use of recycling are just a few examples of how industries are embracing sustainability.

Moreover, the rise of **circular economies**, in which products are designed to be reused, repaired, or recycled at the end of their life cycle, signals a fundamental shift in how we approach production. This transition requires innovative thinking at every stage of the process, from design to disposal. In the coming years, we can expect to see even greater emphasis on creating products that minimize environmental impact while maintaining high levels of performance and functionality.

The Journey Continues

While this book has explored a wide range of fascinating processes and industries, it only scratches the surface of the vast world of manufacturing, engineering, and innovation. Each chapter has offered a glimpse into the hidden stories behind the products and technologies that shape our world, but there is so much more to discover. This is just the beginning of our journey into the world of "how things are made."

As you close the pages of **"Behind the Scenes: Secrets of How Things Are Made – Book One,"** I hope you walk away with a deeper appreciation for the remarkable processes that make up the world around us. Whether you're marveling at the complexity of a modern car engine or simply enjoying a cup of coffee, you now understand that behind every object, big or small, lies a fascinating story of craftsmanship, innovation, and collaboration.

This book may be called "Book One" for a reason: the world is full of untold stories waiting to be explored. The next time you encounter a new product, see a construction site, or hear about a technological breakthrough, take a moment to think about the countless steps, the people, and the expertise involved in bringing that creation to life. There are endless "secrets" left to uncover, and the second volume of this book will continue to delve into the behind-the-scenes processes that fuel our modern world.

Final Thoughts

The world is an incredible place, full of ingenuity, creativity, and perseverance. Behind every building, every technological device, every item of clothing, and every work of art, there is a process—a series of steps that brings an idea to reality. As we continue to innovate and evolve, the methods of how things are made will transform, adapting to new challenges and possibilities.

By uncovering these secrets, we gain not only knowledge but a deeper connection to the world around us. Whether you're an aspiring engineer, a curious mind, or simply someone who loves to learn, the stories within this book remind us that behind every creation, there is an element of wonder. And with that wonder comes the drive to keep exploring, discovering, and creating.

Thank you for joining me on this exploration of **how things are made**. I look forward to continuing this journey with you in future books, as we dive deeper into the fascinating and ever-evolving world of innovation.

Behind the Scenes

Bibliography

1. **"How Things Work: The Inner Life of Everyday Machines"** by Theodore Gray

This visually stunning book explores the mechanics behind everyday objects and machines, offering a perfect introduction to how things are made and how they function.

2. **The Way Things Work Now"** by David Macaulay

A brilliant visual guide to understanding the technology and machines around us, this book dives into the detailed processes behind common inventions and machinery.

3. **The Perfectionists: How Precision Engineers Created the Modern World"** by Simon Winchester

Focuses on the role of precision engineering in the development of everything from clockworks to space travel, showing how meticulous craftsmanship has shaped technology.

4. **How is it Done?** by The Reader's Digest Association Limited.

5. **National Geographic: Ultimate Factories Collection:** DVDs

Behind the Scenes

7. **Paper: Paging Through History** by Mark Kurlansky

A fascinating history of one of humanity's most important materials: paper. It explores how paper is made and its impact on culture, industry, and technology.

8. **How Things Work**

This DK book is a visual guide that covers the mechanisms behind various devices, tools, and systems, from simple machines to complex gadgets.

9. **National Geographic Science of Everything: How Things Work in Our World**

A visually stunning book that explains the science behind everything from biology to technology, showcasing the fascinating processes that shape our everyday lives.

Acknowledgments

Writing **"Behind the Scenes: Secrets of How Things Are Made"** has been an incredible journey of discovery, collaboration, and learning, and it would not have been possible without the support, insights, and encouragement of many individuals.

First and foremost, I would like to express my deepest gratitude to the countless innovators, engineers, scientists, and creative minds whose incredible work has inspired the pages of this book. Your dedication to advancing knowledge, solving problems, and pushing the boundaries of what is possible has provided the foundation for many of the stories told here. Your brilliance and ingenuity continue to amaze me, and this book is a testament to the contributions you have made to shaping our world.

A heartfelt thank you to my family and friends, who have been a constant source of support throughout this project. Your patience, understanding, and words of encouragement have kept me motivated through the many late nights of research and writing. To my spouse and son, your unwavering belief in me has been a guiding light, and I am forever grateful for your love and support.

To my editor, your keen eye, insightful feedback, and tireless efforts have helped shape this book into what it is today. Thank you for pushing me to refine my ideas and for being such an integral part of the creative process. Your expertise and dedication have been invaluable, and I deeply appreciate your partnership on this journey.

I would also like to extend my thanks to the researchers and subject matter experts who generously shared their knowledge and insights. Your contributions helped bring depth and clarity to some of the most complex topics covered in this book. Your willingness to explain intricate processes and break down complicated concepts has enriched my understanding, and I hope this book reflects the brilliance of your work.

A special acknowledgment goes out to my readers—those who are as curious about the world as I am. It is for you that this book was written. I hope it inspires you to look at the everyday world with fresh eyes, sparking the same wonder and amazement I experienced while uncovering the fascinating processes behind the things we often take for granted. Thank you for your interest and for joining me on this exploration.

Finally, I would like to express my gratitude to everyone who played a part in bringing **"Behind the Scenes: Secrets of How Things Are Made"** to life. From the publisher who believed in this project to the countless people behind the scenes who helped turn this vision into reality, I am deeply thankful for your support and collaboration.

This book is a celebration of human creativity and the marvels of innovation, and I am honored to share it with you.

With sincere gratitude,

Zahid Ameer
Versatile Indie Author

Disclaimer

The information provided in this book, *"Behind the Scenes: Secrets of How Things Are Made,"* is for educational and informational purposes only. While every effort has been made to ensure the accuracy and reliability of the content, the author and publisher do not make any representations or warranties of any kind, express or implied, regarding the completeness, accuracy, reliability, suitability, or availability of the information contained within this book. Any reliance you place on such information is strictly at your own risk.

The processes, methods, and technologies described in this book are subject to change due to ongoing advancements in science, engineering, and manufacturing. Readers are encouraged to seek additional professional advice or conduct independent research if using this information for any practical or commercial purpose.

The author and publisher will not be held liable for any errors, omissions, or any outcomes related to the use of this information. Any specific company names, trademarks, or products mentioned in this book are referenced solely for illustrative purposes and do not imply endorsement or affiliation.

Behind the Scenes

By reading this book, you agree to hold the author and publisher harmless for any and all claims or damages arising from the use or misuse of the information contained herein.

About me

I am Zahid Ameer, hailing from the vibrant country of India. As an author, ghostwriter, bibliophile, online affiliate marketer, blogger, YouTuber, graphic designer, and animal lover, I have woven my passions into a unique tapestry that defines my life's work.

Born and raised in India, I have always possessed a deep love for literature. With an insatiable appetite for books, I have amassed an impressive collection of around 1,500 titles, predominantly in English. My passion for reading brings me immense joy and serves as a source of inspiration for my writing endeavors.

I have compiled an impressive portfolio of written works as an author and ghostwriter. With a captivating writing style and an innate ability to craft engaging narratives, I bring my stories to life, captivating readers from all walks of life. My wide range of interests and experiences contribute to the richness of my writing, allowing me to connect with my audience on a heartfelt level effortlessly.

Beyond my literary pursuits, I have also established a strong presence on various digital platforms. I utilize my YouTube channel and blog to raise awareness about all types of knowledge and to share heartwarming stories of animals. Using my platform to shed light on important issues, I strive to create a world where humans and animals can coexist harmoniously.

In addition to my work as an author, I have also dabbled in the world of affiliate marketing. With my webpreneur spirit, I have ventured into online marketing, leveraging my knowledge and skills to promote products and services that align with my values.

However, my most cherished role is that of a father. Family is at the core of my being, and everything I do is centered around creating a better future for my loved ones. My dedication to my family is evident in my passion for personal growth and my relentless pursuit of success. Through my various endeavors, I strive to set an example of perseverance and ambition for my children, inspiring them to chase their dreams unapologetically.

In a world where specialization often dominates, I defy convention by embracing multiple passions and excelling in diverse fields. My love for books, animals, and family has become the driving force behind my achievements. By the grace of Almighty God, my unique blend of characteristics has allowed me to leave an indelible mark on the world, enriching the lives of those I encounter along the way.

To your grand success in life,

Behind the Scenes

Zahid Ameer
Versatile Indie Author

www.ingramcontent.com/pod-product-compliance
Lightning Source LLC
Chambersburg PA
CBHW052138220526
45471CB00004B/1434